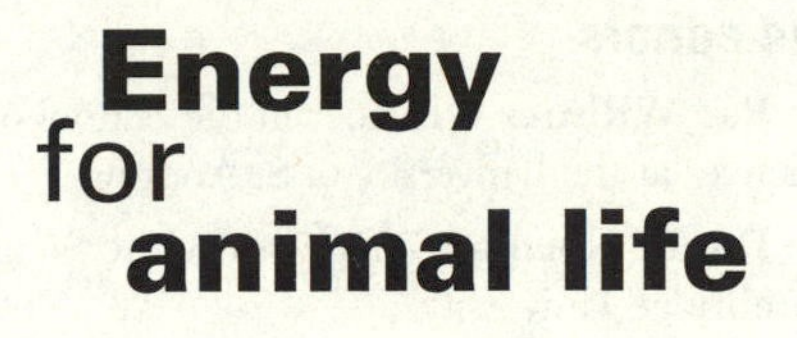

Energy for animal life

Oxford Animal Biology Series

Series editors: Pat Willmer and David Norman

The editors

Dr Pat Willmer is Reader in the School of Biological and Medical Sciences at the University of St Andrews.

Dr David Norman is Director of the Sedgwick Museum at Cambridge University.

Advisers

The role of the advisers is to provide an international panel to help suggest titles and authors, to ensure individual countries' teaching needs are met, and to act as referees.

The aim of the **Oxford Animal Biology Series** is to publish attractive supplementary texts in comparative animal biology for undergraduates studying biological science by adopting a fresh, integrated approach. The series has two distinguishing features. One, that topics within each book are addressed using examples from throughout the animal kingdom, looking for parallels that transcend taxonomy; and two, all aspects of the topics are chosen to match existing and proposed courses and syllabuses, while taking into account the depth of coverage needed and the amount of space available. Further reading sections, consisting mainly of review articles and books, guide the student into the literature at greater depth. The series is international in scope, both in species used as examples and in references to scientific work.

Energy for animal life

R. McNeill Alexander
School of Biology
University of Leeds

OXFORD
UNIVERSITY PRESS

OXFORD
UNIVERSITY PRESS

Great Clarendon Street, Oxford OX2 6DP

Oxford University Press is a department of the University of Oxford
and furthers the University's aim of excellence in research, scholarship,
and education by publishing worldwide in

Oxford New York

Athens Auckland Bangkok Bogotá Buenos Aires Calcutta
Cape Town Chennai Dar es Salaam Delhi Florence Hong Kong Istanbul
Karachi Kuala Lumpur Madrid Melbourne Mexico City Mumbai
Nairobi Paris São Paulo Singapore Taipei Tokyo Toronto Warsaw

and associated companies in Berlin Ibadan

Oxford is a registered trade mark of Oxford University Press

Published in the United States
by Oxford University Press Inc., New York

A catalogue record for this book is available from the British Library

Library of Congress Cataloging in Publication Data
(Data available)

ISBN 0 19 850053 X (Hbk)
ISBN 0 19 850052 1 (Pbk)

Typeset by Footnote Graphics, Warminster, Wilts
Printed in Great Britain by
Biddles Ltd, Guildford and King's Lynn

Preface

Life depends on energy, in most cases on the energy that comes to the earth as solar radiation. Much of the activity of animals is devoted to getting the food which is their energy source. Discussions of the relative merits of different behaviour patterns and life history strategies often pose the question, how can the animal best use its limited resources of energy? Energy is central to our understanding of animal life.

This book is about how animals get energy, and how they use it. It starts with solar radiation and its capture by photosynthesis both on land and in the sea. Next it discusses the energy needed for the basic processes of life, for simply living without doing anything. Then it examines food sources both for herbivores and for carnivores and compares the merits of different designs of digestive system, and of different strategies for finding and choosing food. After that it discusses the energy costs of running, swimming and flight, which represent a large part of the energy budget of many animals. Next, it considers the energy costs of growth and reproduction, discussing how a limited supply of energy can best be used. Then there is a chapter on body temperature, which shows how the processes of life (including energy consumption) go faster at higher temperatures. Some animals are heated by the sun and others (like ourselves) are warmed by their own metabolism. The final chapter looks at the energy budgets of a variety of animals, assessing the energy costs of major activities in everyday life in the wild.

This book is designed principally for first and second year undergraduates, but it seems likely that actual readers will range from sixth-formers to research workers.

Leeds R.McN.A.
February 1998

Contents

1 Energy sources

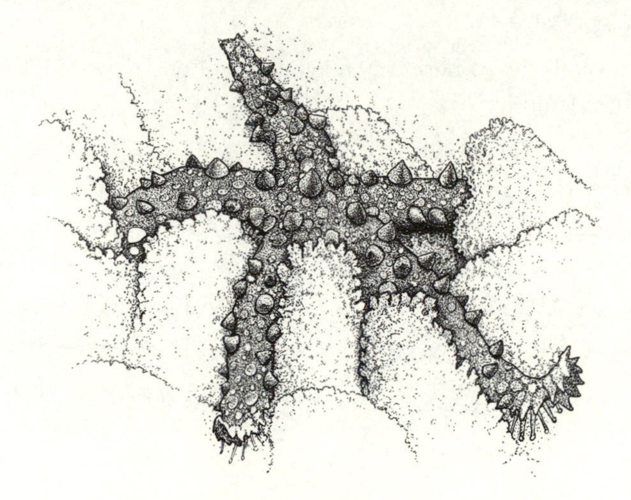

1.1 Energy from the sun

Life on earth is supported by energy from the sun. This chapter is mainly about the sun's radiation and how it is captured by plants. A short section at the end describes an alternative energy source used by a few exceptional organisms.

The light from the sun is its most obvious radiation, but is just part of a broad spectrum. The visible spectrum ranges from violet light, with wavelengths of about 0.4 micrometres (μm), to red, with about 0.7 μm. We cannot see shorter or longer wavelengths. Ultraviolet radiation, with wavelengths less than 0.4 μm, causes suntanning and skin cancer. Infrared radiation, with wavelengths greater than 0.7 μm, is also invisible, but we can feel its warming effect on our skin.

All objects emit radiation. Cold bodies emit only a little, but as you heat a body up it emits more and more radiation, at shorter and shorter wavelengths. A piece of iron at room temperature emits only infrared, with wavelengths much longer than 0.7 μm, but as a blacksmith heats it, its radiation increases and moves into the visible spectrum. A red-hot bar is emitting red light (and plenty of infrared as well). A white-hot bar emits the whole range of visible wavelengths. The sun emits the range of wavelengths that *any* body would emit at its temperature of about 5500°C. About 40 per cent of its energy is conveyed as light (0.4–0.7 μm), 40 per cent as near infrared (0.7–1.5 μm) and the rest as shorter or longer wavelengths. Ultraviolet is known for its harmful effects but amounts to only 8 per cent of the energy.

The earth, rotating around the sun, remains at a near-constant distance from it, so solar radiation reaches the outer layers of the atmosphere at a near-constant rate. To describe this we need some definitions. **Radiant flux** means the rate at

Box 1.1 **Energy and power**

Forces are measured in newtons (abbreviation N), energy in joules (J), and power in watts (W).

One joule equals the work done when a force of one newton moves one metre along its line of action.

One watt is a rate of energy supply or consumption of one joule per second.

As for other units, prefixes are used to indicate multiples or fractions. For example:

one kilonewton (kN) = 10^3 newtons
one megawatt (MW) = 10^6 watts
one gigajoule (GJ) = 10^9 joules
one millijoule (mJ) = 10^{-3} joules
one micronewton (μN) = 10^{-6} newtons
one nanowatt (nW) = 10^{-9} watts
one picowatt (pW) = 10^{-12} watts

which energy is conveyed by radiation. It may be used to describe the rate at which energy is emitted (for example by the sun) or received (for example by the earth). It is measured in watts (see Box 1.1; abbreviation W). **Radiant flux density** means the radiant flux emitted or received by a surface of unit area and is given in watts per square metre ($W\,m^{-2}$). Satellites have been used to measure the radiant flux density just outside the earth's atmosphere, on a surface held at right angles to the sun's rays. It is 1370 $W\,m^{-2}$.

Even on the clearest days, the radiation reaching the ground is much less than this. Maximum values (on surfaces at right angles to the sun) are about 1100 $W\,m^{-2}$ in Israel and 1000 $W\,m^{-2}$ in murkier England. About half of the loss is due to absorption. The ozone in the upper atmosphere absorbs a little visible light and much of the ultraviolet. Other gases, especially water vapour, absorb mainly infrared. The other half of the loss in a cloud-free atmosphere is due to scattering. Gas molecules scatter light in all directions, most effectively at the blue end of the spectrum. This is what makes the sky look blue—light that was not initially travelling in our direction is scattered towards us from all parts of the sky. From rockets in space, the sky looks black. Light is also scattered by particles of dust and smoke, which deflect it by only small angles from its initial direction.

At sunrise and sunset the sun's rays travel obliquely through the atmosphere, making their path through it much longer (Fig. 1.1(a)). So much of the blue light may be scattered that the sun's disc looks red.

Clouds scatter light back towards the sun, which is why clouds that are dull grey from below are brilliant white when seen from an aeroplane above them. Scattering from the top of the cloud and absorption of radiation passing through may reduce the radiant flux density on the ground to 10 per cent of what it would be if the sky were clear. Under thick cloud that hides the sun, direct rays from the

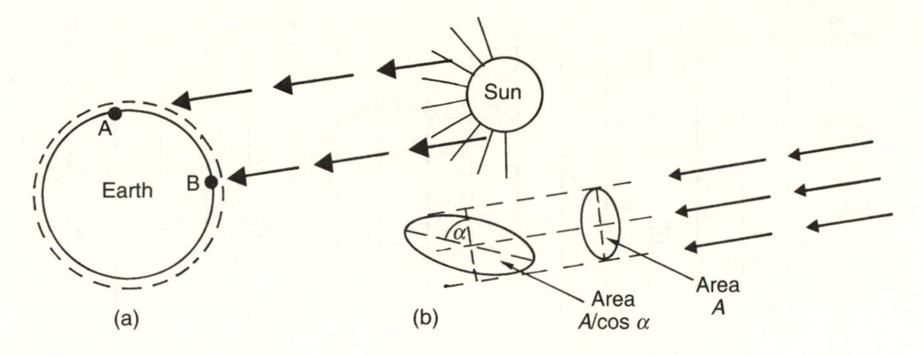

Fig. 1.1 (a) The sun's rays have a longer path through the atmosphere to point A on the earth's surface, where it is dawn, than to point B, where it is noon. (b) Radiant flux density is less on an oblique surface than on one perpendicular to a beam of light.

sun are almost eliminated (so there are no shadows) and almost all the radiation reaching the ground has been scattered.

The flux densities that have been quoted so far have been those falling on surfaces at right angles to the rays. In many cases, the flux density that is important in ecology is that on the surface of the ground, which is generally oblique to the rays. Consider a beam of cross-section A and flux F. The radiant flux density on a surface at right angles to the rays is F/A but on a surface tilted at an angle α (see Fig. 1.1(b)) it is only $(F/A)\cos\alpha$, because the beam is spread out over a larger area. For this reason, radiant flux densities reach higher values in the tropics (where the sun rises higher in the sky) than in temperate latitudes. Also, the sun rises higher in summer than in winter, so flux densities reach higher values in summer. In early morning and late evening, flux densities are low both because the rays reach the ground at a shallow angle and also because of losses on their long path through the atmosphere.

Figure 1.2 shows how radiant flux density on the ground in southern England fluctuated in the course of cloudless days in summer, autumn, and winter. The latitude was 52°N. The earth's axis is inclined at 66.5° to the plane of its orbit, so in the course of a year the latitude at which the sun is vertically overhead at noon fluctuates between (90 – 66.5) = 23.5°N and 23.5°S. At latitude 52°N the angle α (Fig. 1.1(b)) at which level ground is tilted to the sun's rays at noon therefore varies from (52 – 23.5) = 28.5° on the longest day (June 21) to (52 + 23.5) = 75.5° on the shortest (December 21). If the direct radiation from the sun had a flux density of 1000 W m^{-2} on a surface at right angles to it, its values on level ground at noon on these dates would be 1000 cos 28.5° = 879 W m^{-2} and 1000 cos 75.5° = 250 W m^{-2}. The observations in Fig. 1.2 were not made exactly on these dates (cloudless days are rare in England), but close to them.

Figure 1.3 shows radiant flux density in the same place in England on a day of broken cloud. Notice how it falls to about one quarter of the sunlit values when-

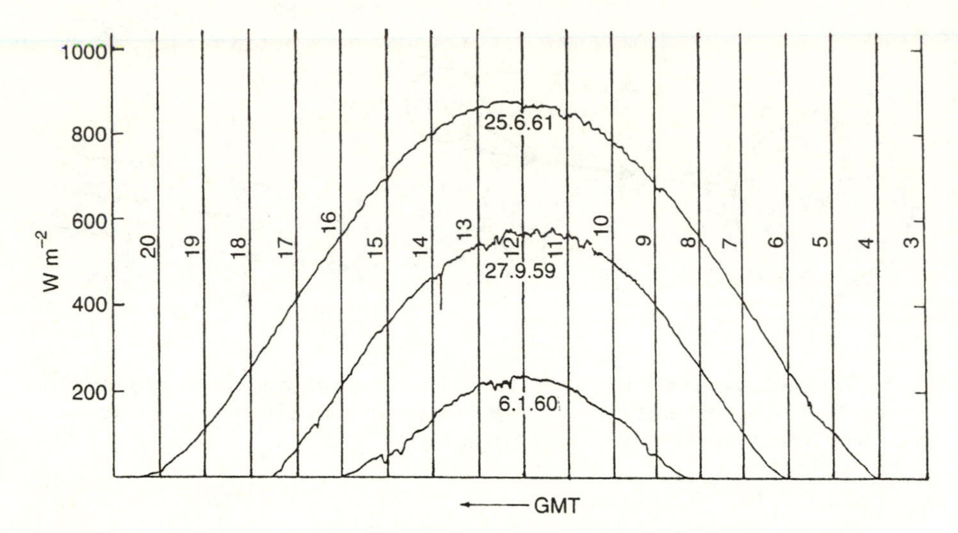

Fig. 1.2 Records of solar radiant flux density on a horizontal surface, for cloudless days in June, September, and January at Rothamsted, England. Note that the time scale (hours) reads from right to left. (From J. L. Monteith and M. H. Unsworth (1990) *Principles of Environmental Physics* (2nd edn). Arnold, London.)

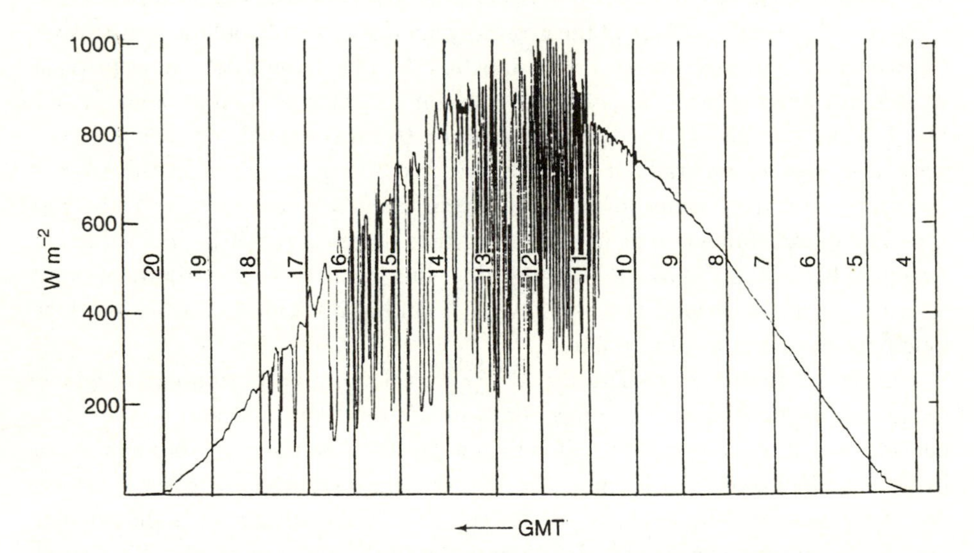

Fig. 1.3 A record of radiant flux density at the same place as in Fig. 1.2, for a June day of broken cloud. (From J. L. Monteith and M. H. Unsworth (1990) *Principles of Environmental Physics* (2nd edn). Arnold, London.)

ever a cloud covers the sun. It rises higher than the cloud-free value immediately before and after each dull period, due to scattering from the edge of the cloud.

The areas under the graphs in Fig. 1.2 represent the quantity of energy arriving in the course of the whole day. It is far less in winter than summer, because of the shorter days and lower flux densities at noon. The energies shown by these three records are about 30, 15, and 4 megajoules per square metre in summer, autumn, and winter. (If you need an explanation of megajoules, see Box 1.1 again.) Remember that Fig. 1.2 refers to cloudless days. Many days are cloudy, and average daily energies for the same place would only be about 16 MJ m^{-2} in summer and 2 MJ m^{-2} in winter.

These are values for dull, cloudy England. Mean daily radiation in different parts of the United States range from about 23 to 31 MJ m^{-2} in summer and from 6 to 12 MJ m^{-2} in winter. This is partly due to lower latitudes and partly to finer weather.

When we discuss the growth of plants, we will be less interested in the amount of energy available in a day than in the amount arriving in a year. In our example of southern England this is 3.2 gigajoules per square metre.

1.2 **Photosynthesis**

Plants capture the sun's energy by the process of photosynthesis, combining carbon dioxide and water to form carbohydrates and oxygen.

$$CO_2 + H_2O + \text{about 8 photons} \Rightarrow CH_2O + O_2 \quad (1.1)$$

In this equation, CH_2O represents a carbohydrate; for example, glucose is $C_6H_{12}O_6$, six times the unit represented in the equation. The photons (particles of light energy) supply the energy needed for the conversion.

The light is captured by pigments in the plant cells, the chlorophylls, together with smaller quantities of carotenoids and (in red algae and cyanobacteria) the phycobilins. These pigments absorb light strongly both near the red end and near the blue end of the visible spectrum, but absorb intermediate wavelengths less strongly; leaves look green because a lot of the green light that falls on them is reflected rather than absorbed. The range of wavelengths that can be used in photosynthesis (photosynthetically active radiation or PAR) is more or less the same as the range that is visible to our eyes. The approximately half of the radiant energy arriving at the earth's surface, which falls outside this range, cannot be used.

Photons of different colours of light have different quantities of energy; the shorter the wavelength, the higher the energy. Thus a mole of red photons has about 180 kilojoules of energy and a mole of blue photons about 300 kilojoules (see Box 1.2). We will soon do a rough calculation for which we will need an average value for the energy content of a mole of PAR: 230 kilojoules.

Whereas photosynthesis captures the sun's energy in foodstuffs, respiration releases the energy again, making it available for the processes of life. It is in effect the reverse of photosynthesis. Respiration of carbohydrate goes like this:

$$CH_2O + O_2 \Rightarrow CO_2 + H_2O + 468 \text{ kJ} \quad (1.2)$$

The energy shown in this equation, 468 kilojoules, is that released when one mole of CH_2O (30 grams) is oxidized.

Box 1.2 **Moles**

The masses of atoms and molecules are generally given as multiples of the mass of a hydrogen atom. Examples of atomic masses are: hydrogen, 1; carbon, 12; oxygen, 16. Thus the molecular mass of carbon dioxide, CO_2, is $12 + (2 \times 16) = 44$.

One mole of a substance is the number of grams equal to the molecular mass, 44 g in the case of carbon dioxide.

A mole contains 6.0×10^{23} molecules.

A mole of photons is 6.0×10^{23} photons.

Comparing our equations for photosynthesis and respiration, we see that 8 moles of photons are needed to make carbohydrate from which 468 kilojoules of energy can be released by respiration. Those 8 moles carry an average of 230 kilojoules of energy each, 1840 kilojoules in all. Only half the energy of solar radiation is in the photosynthetically active range of wavelengths, so 3680 kilojoules of solar energy must arrive on earth to supply 8 moles of PAR and store 468 kilojoules as carbohydrate. If that were the whole story, the efficiency of the process would be $468/3680 = 0.13$. We will soon see that plants capture even smaller fractions of solar energy than this calculation suggests.

1.3 **Primary production**

The rate of photosynthesis, whether in crops or in natural ecosystems, is described as primary production ('primary' to distinguish it from the secondary production that occurs when an animal eats a plant or another animal and converts this food to its own tissues). It can be expressed either in terms of the quantity of energy captured or of the mass of new material formed from carbon dioxide. Gross primary production describes the rate at which photosynthesis is occurring; net primary production describes the rate at which energy or matter is being accumulated as growth or food stores. Net primary production equals gross primary production minus respiration and, for many ecosystems, is between 20 per cent and 70 per cent of gross primary production. Because this book focuses on animals, thinking of plants mainly as potential food, we will be considering net primary production. That will be convenient, because net production is much the easier quantity to measure. To measure gross production you must perform a sophisticated experiment, but to measure net production you have only to harvest samples of your crop from time to time and find out how much its mass or energy content has increased.

Primary production is generally given as a rate per unit area of land. There is generally a growing season in which it is faster than during the rest of the year.

The highest rates observed in crops over periods of a few weeks are around 50 grams dry mass per square metre per day (rates are generally given in terms of dry mass because that is a much better indicator of energy content than the fresh mass including water). For example, in one trial Sudan grass (*Sorghum vulgare*) in California grew at a rate (net primary production) of 51 g m^{-2} d^{-1}, over a period of 35 days. Other studies recorded sugar cane growing in Hawaii at 37 g m^{-2} d^{-1} over a period of 90 days (an unusually long period for such fast growth to be sustained) and rice growing in Japan at 36 g m^{-2} d^{-1} over 21 days.

We will convert these values to energy so that we can compare the energy captured by photosynthesis to the energy that was available from solar radiation. The equation for respiration (eqn 1.2) showed that one mole of CH_2O (30 g) releases 468 kJ when oxidized, 15.6 kilojoules per gram. Corresponding data for proteins and fats are about 24 and 39 kJ g^{-1} respectively. In the light of these data, it is not surprising to find that dried plant tissues generally yield about 17.5 kJ g^{-1} when oxidized. Thus net primary production was $51 \times 17.5 = 890$ kJ m^{-2} d^{-1} in the case of the grass and 650 and 630 kJ m^{-2} d^{-1} in those of the sugar cane and rice.

Solar radiation on the crops during the period of the observations averaged 29 MJ m^{-2} d^{-1} in the case of the grass, 17 MJ m^{-2} d^{-1} in the case of the sugar cane, and 20 MJ m^{-2} d^{-1} in the case of the rice. Thus the fractions of solar energy captured by photosynthesis were $0.89/29 = 0.03$ in the case of the grass (remember that 890 kJ = 0.89 MJ), $0.65/17 = 0.04$ in that of the sugar cane, and $0.63/20 = 0.03$ in that of the rice. These fractions are far below our theoretical estimate (in Section 1.2) that it should be possible for plants to capture 0.13 of solar radiation. Some of the difference is due to the energy used in the plant's metabolism; remember that our calculations were based on net, not gross, production.

We are thinking of plants as sources of energy for animals, so we may be more interested in production averaged over the whole year rather than in peak values at the height of the growing season. Intensively managed grasslands in Britain may produce up to 2.0 kg dry mass m^{-2} a^{-1} ('a', representing the Latin word *annus*, is the internationally accepted abbreviation for 'year'). This corresponds in terms of energy to about 35 MJ m^{-2} a^{-1}. Solar radiation in southern England amounts to about 3200 MJ m^{-2} a^{-1}, so about 1 per cent of the annual radiation is captured. Grasslands in New Zealand may produce up to 60 MJ m^{-2} a^{-1}, and sugar cane plantations and young tropical forests may each produce up to 120 MJ m^{-2} a^{-1}. These data refer to sunnier climates than Britain, but still show more efficient energy capture than the British grassland.

Fractional energy capture on land may seem low, but in water it is generally lower. Plankton production cannot be measured like production of land plants, by harvesting the crop at the end of the season, because each individual organism lives only for a short time. Instead, it must be estimated indirectly, for example by measuring growth rates in laboratory experiments at appropriate temperatures.

Primary production in the sea is in most places 0.2 kg dry organic matter m^{-2} a^{-1} or less in the open oceans, and 0.5 kg m^{-2} a^{-1} or less in coastal waters.

1.4 Limits to photosynthesis on land

One reason for photosynthesis being so inefficient is the phenomenon of photorespiration: a proportion of the products of photosynthesis is broken down again, using oxygen and releasing carbon dioxide. Energy previously captured by photosynthesis is thereby lost; unlike true respiration, photorespiration does not produce ATP or similar compounds which the plant could use as energy sources.

There are two alternative processes of photosynthesis, only one of which is accompanied by photorespiration. Most plants use only the C_3 process, so called because the first product of fixing carbon dioxide is a compound with three carbon atoms in its molecule. One molecule of a sugar with five carbon atoms (ribulose 1,5-bisphosphate, abbreviated as RuBP) combines with one molecule of carbon dioxide to produce two molecules of the three-carbon compound glyceraldehyde 3-phosphate (PGA).

$$\underset{\text{5 carbons}}{\text{RuBP}} + CO_2 + H_2O \Rightarrow \underset{\text{3 carbons}}{\text{2 PGA}} \tag{1.3}$$

This reaction is catalysed by the enzyme known by the abbreviated name of RUBISCO, which makes up more than 15 per cent of the protein in chloroplasts and has been claimed to be the most abundant protein in the world. Unfortunately, RUBISCO also catalyses another reaction that produces glycollate (hydroxyethanoate), a two-carbon compound:

$$\underset{\text{5 carbons}}{\text{RuBP}} \Rightarrow \underset{\text{3 carbons}}{\text{PGA}} + \underset{\text{2 carbons}}{\text{glycollate}} \tag{1.4}$$

This reaction is followed by partial oxidization of the glycollate, the process of photorespiration. Some of the glycollate carbon atoms are reconverted to useful compounds but some are lost as carbon dioxide. When carbon dioxide is plentiful and oxygen sparse, reaction 1.3 is favoured, but when oxygen is plentiful and carbon dioxide sparse, reaction 1.4 proceeds rapidly. Photorespiration can be inhibited experimentally by reducing the oxygen content of the atmosphere to below 2 per cent (the normal value is 21 per cent). This increases the net rate of photosynthesis of plants that depend on the C_3 process by up to 50 per cent.

In some flowering plants, including many grasses, a C_4 photosynthetic process operates as a preliminary to the C_3 process. Phosphoenolpyruvate (three carbons) reacts with carbon dioxide to give oxaloacetate (four carbons; hence the name C_4) and phosphate. This happens in mesophyll cells, but in C_4 plants nearly all the RUBISCO is in specialized cells (the Kranz cells) which are clustered around the vascular bundles. The oxaloacetate produced by the C_4 process is immediately converted to malate (hydroxybutanedioic acid) or aspartate (aminobutanedioic acid), which are transported to the Kranz cells, where the captured carbon dioxide is released again, to be recaptured by the C_3 process. The advantage of this is that a high concentration of carbon dioxide is maintained in the Kranz cells, almost eliminating photorespiration. The accompanying disadvantage is that energy is needed (as ATP) to drive the C_4 process. The balance of advantage tends to favour C_4 plants in hot, dry climates and C_3 plants elsewhere, but neither C_3 nor C_4 plants can use solar energy with the efficiency of 13 per cent that our initial discussion suggested. Among the crops for which we estimated efficiencies of 3 per cent to 4 per cent, rice is a C_3 plant while Sudan grass and sugar cane are C_4 plants.

These considerations alone are not enough to explain the discrepancy between observed efficiencies and our initial estimate. In addition, leaves reflect 10 per cent or so of the light that falls on them (if they did not reflect they would look black) and some light falls on the ground without being intercepted by leaves. More important, leaves in bright sunlight cannot make full use of the energy of the light that falls on them because the rate of photosynthesis is limited by the scarcity of carbon dioxide in the atmosphere (which contains only 0.03 per cent of carbon dioxide by volume). The limiting effect of the carbon dioxide supply has been demonstrated by showing that plants grow faster in atmospheres with increased carbon dioxide.

The rate of photosynthesis P at different radiant flux densities Φ can be described by the equation

$$P = P_{max}\Phi/(K + \Phi) \tag{1.5}$$

where K is a constant. This is a Michaelis–Menten equation of the kind that applies very generally to enzyme-catalysed reactions. In dim light, when Φ is small compared with K, $P \approx P_{max}\Phi/K$, so the efficiency is P_{max}/K. As Φ increases efficiency falls and, however bright the light, P can never exceed P_{max}. However, P_{max} is not constant; it depends on the temperature (it is greatest in moderately warm conditions) and on the carbon dioxide concentration in the leaf (the greater the concentration, the greater is P_{max}).

We will use eqn 1.5 to make some simple calculations about photosynthesis in bright sunlight. Figure 1.4 has been drawn from the equation with $P_{max} = 25\ \mathrm{W\ m^{-2}}$ and $K = 200\ \mathrm{W\ m^{-2}}$; these values have been chosen to give a graph like those obtained in experiments, when plants have been grown in light of different intensities. Note that they make P_{max}/K (the maximum efficiency in dim light) approximately equal to our theoretical value of 0.13. However, in bright sunlight with $\Phi = 1000\ \mathrm{W\ m^{-2}}$ the rate of photosynthesis is $21\ \mathrm{W\ m^{-2}}$ and the efficiency is only 0.02.

Now we will use Fig. 1.4 to ask, how fast can photosynthesis be in sunny places? To make things simple, we will assume that the sun shines vertically downward with radiant flux density $1000\ \mathrm{W\ m^{-2}}$. You might suppose that the best strategy

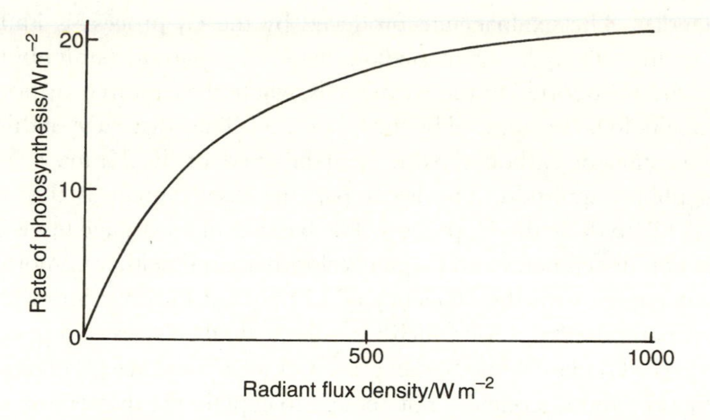

Fig. 1.4 A graph of rate of photosynthesis against radiant flux density, calculated from eqn 1.5.

for a plant would be to cover the ground completely with a single layer of horizontal leaves. In that case, every leaf would receive 1000 W m^{-2} and photosynthesise at a rate of 21 W m^{-2}. Leaf area would equal ground area and the rate of photosynthesis would be 21 watts per square metre of ground.

Now suppose instead that every leaf sloped at 20° to the vertical (Fig. 1.5(a)). Because the leaf is at an angle, each beam of light is spread out over a larger area on its surface and the radiant flux density on the leaves is only (1000 sin 20°) = 342 W m^{-2}. Equation 1.5 tells us that this will give 15.8 watts photosynthesis per square metre of leaf surface. This is less than the 21 W m^{-2} for horizontal leaves, but if the tilted leaves completely shade the ground there will be (1/sin 20°) = 2.9 square metres of leaf per square metre of ground and the rate of photosynthesis *per square metre of ground* will be 46 W, more than double the value for horizontal leaves.

Some plants such as grasses have steeply sloping leaves. An alternative strategy seen in many trees is to have a number of incomplete layers of leaves, each letting some of the light through to the layer below. Figure 1.5(b) shows how this might be done. Every leaf in the top layer is in full sunlight. You might suppose that in the second layer, every point on each leaf would either be in full sunlight or in deep shade, but this is not the case. The reason is that the sun's rays are not precisely parallel; light from the left and right edges of its disc reach the earth at slightly different angles. Below each leaf there is an umbra of deep shade which receives no direct rays from the sun and a penumbra of partial shade which receives rays from only parts of the sun's disc. At about 70 leaf diameters below a layer of leaves, the umbra disappears. Below that, radiant flux density is fairly uniform but dimmer than in full sun. Each layer of leaves reduces the flux density further. Let us suppose that leaves are totally opaque (which is not quite true) and that each layer of horizontal leaves covers half the ground area. Then each layer halves the flux density and, if there are six layers, we can calculate the rate of

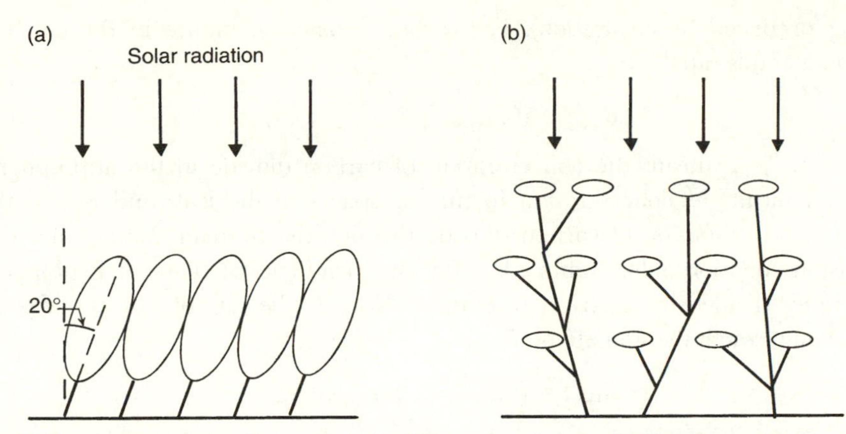

Fig. 1.5 Two arrangements of leaves that give higher rates of photosynthesis than a single layer of horizontal leaves.

photosynthesis as shown in Table 1.1. The total rate of photosynthesis (36 W m^{-2}) is much higher than for a single layer of horizontal leaves and gives a realistic efficiency of 0.036.

Such high rates of photosynthesis are attained in favourable circumstances, but large parts of the earth's surface are much less productive. In many places, this can be explained by water shortage. The problem is that the wider the leaves' stomata open to let carbon dioxide diffuse in for photosynthesis, the faster will water diffuse out and be lost by evaporation into the atmosphere. If water is lost too fast, the plant may be unable to take up water fast enough to replace it, and the plant will wilt.

Let the plant's rate of uptake of carbon dioxide be R_{CO2} (this is the rate at which carbon dioxide is being used by photosynthesis, minus the rate at which it is

Table 1.1 A series of layers of leaves (as in Figure 1.5b) can capture more energy by photosynthesis than a single continuous layer could do. In this hypothetical example, each layer covers half the ground area and so halves the radiant flux density, and photosynthesises at the rate shown in Figure 1.4.

Layer	Flux density Φ (W m^{-2})	Photosynthesis per unit leaf area (W m^{-2})	Photosynthesis per unit ground area (W m^{-2})
1	1000	20.8	10.4
2	500	17.9	8.9
3	250	13.9	6.9
4	125	9.6	4.8
5	63	6.0	3.0
6	31	3.4	1.7
		Total	**35.7**

being produced by respiration). For carbon dioxide to diffuse in through the stomata at this rate

$$R_{CO_2} = (C_{CO_2,atm} - C_{CO_2,leaf})/r_{CO_2} \quad (1.6)$$

where $C_{CO_2,atm}$ means the concentration of carbon dioxide in the atmosphere, $C_{CO_2,leaf}$ means its concentration in the air spaces in the leaf, and r_{CO_2} is the resistance to diffusion of carbon dioxide through the stomata. Notice that this diffusion equation is like Ohm's law for electricity (electric current is potential difference divided by electrical resistance). Now let the rate of loss of water by diffusion through the stomata be R_{H_2O}.

$$R_{H_2O} = (C_{H_2O,leaf} - C_{H_2O,air})/r_{H_2O} \quad (1.7)$$

where the subscripts H_2O indicate that this time the concentrations and resistance refer to water. The carbon dioxide and the water are diffusing in opposite directions along the same path. Big molecules diffuse through air less readily than small ones, and for that reason the resistance of any path to diffusion of carbon dioxide is 1.6 times its resistance to water vapour, so eqns 1.6 and 1.7 give

$$R_{H_2O}/R_{CO_2} = 1.6(C_{H_2O,leaf} - C_{H_2O,air})/(C_{CO_2,air} - C_{CO_2,leaf}) \quad (1.8)$$

A leaf in full sunlight is likely to be hot, often 35°C or more in temperate climates. The air in the spaces within it will be saturated with water vapour and, at 35°C, will contain 40 grams of water vapour per cubic metre. The air outside the leaf will be less moist and, in a dry climate, might contain only 10 g m^{-3} at the same temperature. In this case the concentration difference driving water diffusion would be 30 g m^{-3}. The concentration difference driving carbon dioxide diffusion would be far less and could not exceed 1 g m^{-3}, the concentration of the gas in the atmosphere. Thus R_{H_2O}/R_{CO_2} would have to be at least $1.6 \times 30/1 = 48$; for every gram of carbon dioxide that diffuses into the plant, almost 50 g of water vapour would inevitably be lost. This is the minimum possible water loss. Actual losses are generally much greater, typically about 300 g for C_4 plants and 500 g for C_3 plants. The inevitable loss would be different for different temperatures and humidities but the general principle remains; taking in a little carbon dioxide means losing a lot of water.

If a lot of water is lost from the leaves, water to replace it must be drawn up rapidly from the roots and drawn quickly through the xylem. This requires a large difference between the pressure p_{root} in the xylem in the roots and the pressure p_{leaf} in the xylem in the leaves. The pressure p_{root} must be low enough to draw water out of the soil, as low as minus 80 atmospheres in some desert plants, and p_{leaf} must be even lower. This may present a problem. Leaves depend on the pressure of water in their cells to stiffen them. (Similarly the 'bouncy castles' that children play on are stiffened by the pressure of the air inside.) The osmotic potential of the cell sap makes it possible for the pressure in the cells to be kept positive, although the pressure in the xylem is negative. But if p_{leaf} falls too low, the pressure in leaf cells cannot be maintained, and the leaves wilt. To prevent wilting in dry conditions, the plant must partially close its stomata, reducing the rates of water loss and carbon dioxide uptake.

Cacti and other succulent plants avoid this problem by the extraordinary stratagem of taking their carbon dioxide in at night, when darkness makes photosynthesis impossible. They are enabled to do this by crassulacean acid metabolism, a variant of C_4 photosynthesis in which the C_4 and C_3 processes occur at different times. Carbon dioxide taken in at night by the C_4 process is converted to malic acid and stored until daylight, when the carbon dioxide is released and used for photosynthesis by the C_3 process. The stomata are kept closed during the day, when water would be lost rapidly if they were open. They open at night when the temperature is lower and the air in the spaces in the leaves contains less water. For example, saturated air at 11°C contains only 10 g m^{-3} water vapour, one quarter as much as at 35°C. Also, relative humidity of the air around the leaf will generally be higher at night. Thus the concentration difference driving water vapour loss is much less.

Estimates of primary production in deserts include values of around 3 MJ m^{-2} a^{-1} for grasses in the Sonoran and Great Basin deserts of North America, only one tenth as much as in good agricultural grasslands. *Agave*, a spiky-leaved succulent, uses crassulacean acid metabolism and achieves productions around 10 MJ m^{-2} a^{-1} in the Sonoran desert. It has been found to lose only 25 g water for every gram of carbon dioxide it uses in photosynthesis. Most plants that do not use crassulacean acid metabolism would lose water at least 10 times as fast.

1.5 Limits to plankton production

So far we have considered only production on land, but two thirds of the earth's surface is covered by water in which primary production also occurs. Part of this is in multicellular plants (water weeds and seaweeds) but most is in unicellular organisms, especially plankton which live suspended in the surface waters of both lakes and seas. This section is about marine plankton (but the same principles apply in fresh water). Subsequent sections are about primary production by unicellular symbionts within the cells of multicellular marine animals.

The photosynthetic members of the marine plankton are diatoms, flagellates, and cyanobacteria, all of them unicellular organisms of a few micrometres or tens of micrometres in diameter (Fig. 1.6). The cyanobacteria (formerly called blue-green algae) are prokaryotes, with no nucleus or other membrane-enclosed organelles. The others are eukaryotes. Among the flagellates, the dinoflagellates are particularly important; their characteristics include two flagella, one of them in a groove that encircles the body, and protective plates of cellulose. Curiously, the plates are not external to the cell membrane like the cellulose cell walls of typical plants, but enclosed just below the cell surface in membrane-lined cavities.

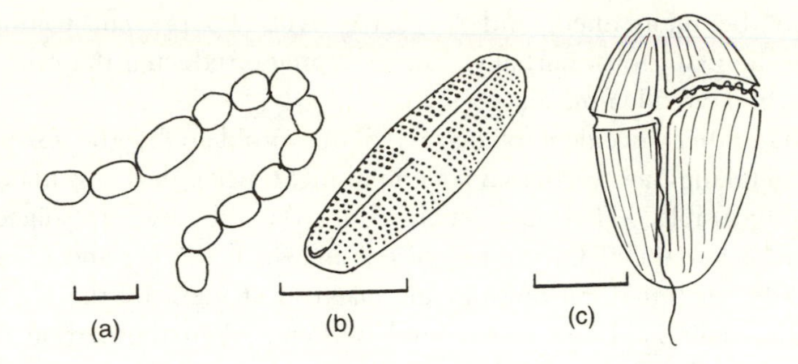

Fig. 1.6 Sketches of photosynthetic members of the plankton: (a) is a cyanobacterium; (b) is a diatom (*Navicula*); and (c) is a dinoflagellate (*Gymnodinium*). The scale bars are 10 μm long.

The deeper you go in the sea, the darker it gets; light is absorbed much more strongly by water than by air. Below a certain depth, daylight is so dim that photosynthetic organisms cannot flourish. As a rough general rule, the radiant flux density must be at least 1 per cent of surface values for net photosynthesis to occur (that is for organisms to be able to take up carbon dioxide by photosynthesis faster than they produce it by respiration). This is similar to the situation for shade-tolerant land plants.

Think of a uniform fluid, illuminated from above. If the radiant flux density is Φ_0 at the surface it is Φ_d at depth d:

$$\Phi_d = \Phi_0 e^{-\alpha d} \quad (1.9)$$

where e is the constant 2.72 and α is the attenuation coefficient for the particular fluid. This coefficient is different for different wavelengths of radiation; in water it is higher for red light than for green and blue, which is why the light deep in the sea is blue-green. The values that follow are rough average values for the whole of the visible (and photosynthetically active) spectrum.

In the clearest oceanic water, the attenuation coefficient is about 0.03 m^{-1} and flux density is about 1 per cent of the surface value at 150 metres ($2.72^{-150 \times 0.03} = 0.01$). Coastal waters are more turbid, with attenuation coefficients typically around 0.3 m^{-1}, giving 1 per cent of surface illumination at 15 metres. Thus the greatest depth at which photosynthesizing plankton can flourish varies greatly from place to place but is generally between 15 and 150 metres.

Light limits the range of depths inhabited by plankton, but it is not the factor that limits production per unit area of the sea surface, any more than it is on land. We saw that even the most productive crops on land capture no more than about 4 per cent of the solar energy that falls on them, far less than seemed theoretically possible; their production seemed actually to be limited by the carbon dioxide supply. We will see that plankton production is generally limited by the supply of nutrients, especially nitrates (which are needed for producing proteins) and phosphates (needed for the phospholipids of cell membranes and for ATP, etc.).

Dramatic evidence of this is provided by lakes polluted by sewage or by fertilizers from neighbouring agricultural land. Lakes that were initially oligotrophic, with low plankton production and clear water, become eutrophic, with high production, and the rich crop of plankton turns their water green. Typical rates of plankton production are of the order of 100 grams dry matter per square metre per year in oligotrophic lakes and 1 kg m^{-2} a^{-1} in eutrophic ones. The latter is similar to production rates for temperate grasslands.

In the sea, it is generally nitrate that limits production. There is always plenty of it deep down in the water; it is only near the surface where the plankton live that it runs short. To understand this we need to know something of the processes that mix water from different depths in the sea, or prevent it from being mixed.

Deep in the oceans of all parts of the world the water is cold, about 2°C. Surface water, however, is heated by the sun, to about 28°C in the tropics. The warm water is less dense than the cold, so tends to remain at the surface. However, wind blowing over the surface sets up swirling eddies in the water, mixing the upper layers of the sea so that its temperature is fairly uniform, down to a certain depth. These two effects, acting together, can produce a well-mixed layer of warm water floating on the deeper cold water. The intermediate layer in which the temperature changes rapidly with depth is described as a thermocline.

In the tropics, where temperature changes little with the seasons, the oceans have permanent thermoclines at depths of a few hundred metres. All the photosynthetic plankton live above the thermocline; below it, it is too dark. There is very little nitrate or phosphate in the water above the thermocline because the plankton have removed the rest. Below the thermocline, where there are no plankton, these ions are more plentiful. Production is limited by the rate at which they become available above the thermocline as dead plankton decay. Some plankton sink through the thermocline and will die in the darkness below; their nutrient content is lost, but the loss is made good by the slight mixing that occurs across the thermocline. Primary production continues all year and typically amounts to around 100 grams dry matter per square metre per year.

Temperate oceans are generally at least as productive as tropical ones, despite lack of production in winter. Primary production around 200 g m^{-2} a^{-1} is common. During the winter, surface water cools to lower temperatures than the water immediately below it. The cold surface water sinks, mixing water of different temperatures and destroying the thermocline that is present only in summer (Fig. 1.7(a)). With no thermocline, stormy weather easily mixes the water, down to substantial depths. Nutrients from deeper water are brought to the surface but there is very little primary production because the water is mixed to so great a depth that plankton are carried down to depths where it is too dark for them to flourish, and the plankton population is depleted. Reproduction in the short winter days cannot keep pace with these losses. In the spring the surface water is warmed and a thermocline forms. It acts as a barrier to mixing; it is harder to mix the less dense warm water into the denser cold water than it would be to mix water of uniform temperature. The danger to plankton of being carried down into the dark depths is reduced, just at the time when warmer water and longer daylight are making conditions more favourable for growth and reproduction. The population of photosynthetic plankton rises dramatically (Fig. 1.8), using most of the available nutrients. Nitrate concentrations in the water above the thermocline fall to low levels (Fig. 1.7(b)). The increase of the photosynthetic plankton (phytoplankton) is closely followed by an equally dramatic increase of the non-photosynthesizing plankton (zooplankton) which feed on them (Fig. 1.8). Nutrient shortage and predation then cause a sharp fall in the phytoplankton population, which remains relatively low through the summer. Zooplankton numbers fall correspondingly. Phytoplankton increases again in autumn (but not to its spring levels), by which time decay of dead plankton and water mixing across the thermocline have somewhat increased nutrient concentrations, and reduced zooplankton populations mean less predation. Then winter comes, the surface water cools, the thermocline breaks down, and the cycle is repeated.

The concentration of nitrogen (as nitrate) near the surface of temperate seas is restored by winter mixing, typically to around 100 milligrams per cubic metre. After the thermocline has formed, the nitrate concentration above it falls as the plankton grow and use nitrate to make protein. Production is limited by the amount of nitrate trapped above the thermocline.

Suppose that the thermocline is at a depth of 50 metres, a typical offshore value for the north Atlantic. That means that between the thermocline and the surface there are 50 cubic metres of water for every square metre of surface. When the concentration of nitrogen (as nitrate) is 100 mg m^{-3}, there are 5 grams of available nitrogen below each square metre of surface. Five grams of nitrogen is enough to

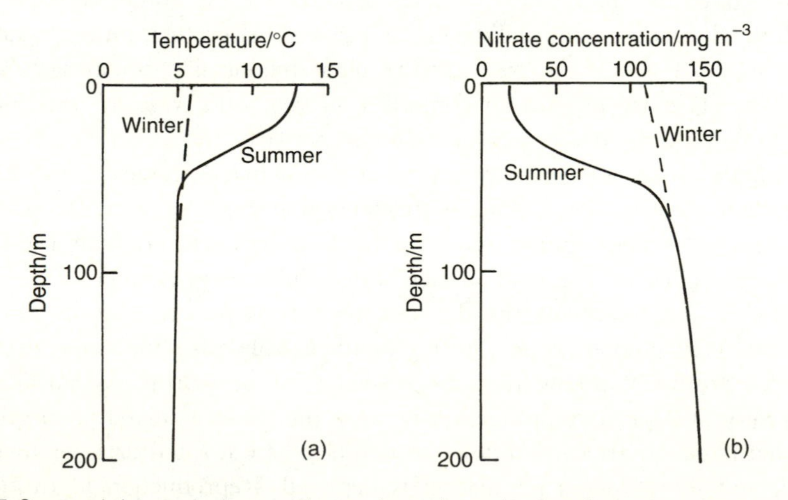

Fig. 1.7 Seasonal changes in the sea. These schematic graphs show how temperature and nitrate concentration (expressed as mg N per cubic metre) change with depth, in summer and winter. The scales are based on observations in the Gulf of Maine. (N. W. Rakestraw (1936) *Biological Bulletin* **71**, 133–167.)

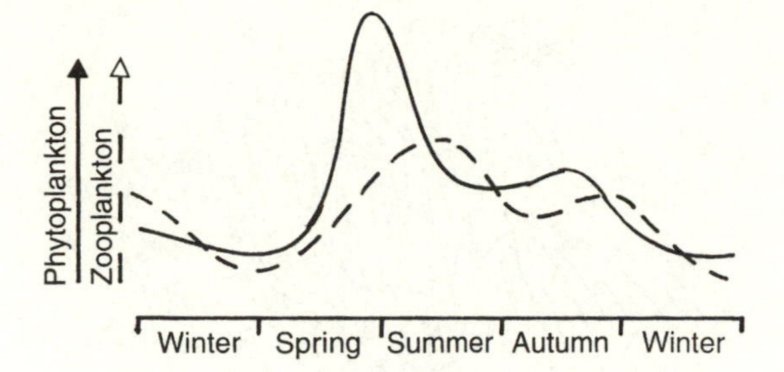

Fig. 1.8 A schematic graph showing how plankton populations change during a year, in a temperate sea. The continuous line refers to phytoplankton and the broken one to zooplankton.

make about 30 grams of protein. The organic matter of plankton is not all protein, but includes fat, carbohydrate, and other nitrogen-free compounds, so 30 grams of protein will be enough for about 100 grams (dry mass) of plankton. This rough calculation suggests that the population density of oceanic plankton cannot exceed 100 g m^{-2}. Annual production may be more than this if dead plankton and the faeces of predators that eat them decay fast enough for their nitrogen to be recycled during a season, but this limit helps us to understand why rates of production are seldom more than about 200 g m^{-2} a^{-1}.

Conditions are more favourable along some coasts where winds blow surface water away from the shore and water rises from deeper to replace it. Such upwellings bring nutrients to the surface, making primary productivities up to 1000 g m^{-2} a^{-1} possible.

1.6 Production in coral reefs

Corals are animals, but the ones that form reefs get most of their energy from photosynthesis. The corals themselves have no chloroplasts, but unicellular algae living within their cells do, and some of the products of their photosynthesis become available to the coral. Because they depend on light, reef-building corals flourish only near the surface, generally at depths of 30 metres or less. Atolls are coral islands that are sinking into the sea floor. Dead coral may extend down to depths of 1000 metres or more, having died as it sank too deep for photosynthesis, but living coral is found only near the surface.

Coral polyps resemble tiny sea anemones projecting from the rock-like coral skeleton (Fig. 1.9). Like sea anemones (to which they are closely related) they have two layers of cells, an outer epidermis and an inner gastrodermis, enclosing the cavity that serves as a gut. The photosynthetic algae are generally present only in the gastrodermis but may be so plentiful as to contain 50 per cent of the coral's protein. They are oval, about 10 μm long, with thick cell walls.

Both the coral and the algal cells can survive alone, without the other. Pieces of living coral can be rid of their algal cells by keeping them in darkness, making

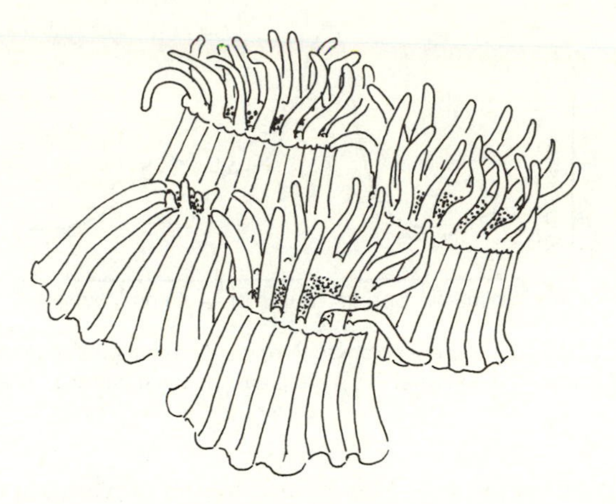

Fig. 1.9 A sketch of part of the surface of a reef-building coral. Polyps of around 5 mm diameter protrude from the calcareous skeleton that they produce.

photosynthesis impossible. The algae eventually degenerate and are expelled from the coral cells. After this, the corals can still survive, if provided with food. On the other hand, the algae can be separated out by grinding up the coral and centrifuging; they will then live and grow. They must be exposed to light and kept in a suitable solution of inorganic salts with traces of vitamins B_1 and B_{12}. When kept in this way, some of the cells change shape and develop flagella, and it becomes apparent that they are dinoflagellates closely related to species of *Gymnodinium* that live free in the plankton (Fig. 1.6(c)).

Pieces of coral kept in daylight in sea water without any food have survived for over a year, suggesting that the animal is being supported by the photosynthesis of the algae. This has been confirmed by experiments in which corals have been kept (again in daylight) in water containing radioactive carbon dioxide, $^{14}CO_2$. After a while, samples of the coral were washed, fixed, and sectioned for microscopy. Autoradiography was used (laying photographic film over the microscope sections) to find radioactivity in the sections. It was found not only in the algal cells, but also in the cells of the coral itself, including the alga-free epidermal cells. The radioactive atoms must have been captured by photosynthesis in the algal cells and passed to the coral.

One possibility was that the coral periodically digested some of the algal cells, but no evidence of that has been seen. Instead, it seems that products of photosynthesis diffuse out of the algal cells into the cytoplasm of the cells that contain them. Evidence for this comes from the observation that when the algal cells are kept in culture, glycerol and other organic molecules diffuse out into the water around them.

A more subtle experiment seems to show that products of photosynthesis leak out of the algal cells, not only in culture but also *in situ* in the coral. Pieces of coral were kept in the light, in sea water containing radioactive sodium bicarbonate,

$NaH^{14}CO_3$. After some time, samples of the water were treated with acid to drive off the $^{14}CO_2$ from the bicarbonate, then analysed for radioactive organic compounds. Virtually none were found; if the algal cells were releasing products of photosynthesis, they were being retained by the coral cells. The experiment was repeated, adding non-radioactive organic compounds as well as radioactive bicarbonate to the water. When glucose was added, [^{14}C]glucose was found in the water at the end of the experiment. When glycerol or the amino acid alanine were added, the radioactive forms of these compounds accumulated in the water. The interpretation of these results is that glucose, glycerol, and alanine made by photosynthesis were all diffusing out of the algal cells. Normally, they were retained by the coral cells, but if one of the compounds had been added to the water there was so much of that compound around that the coral could not use it all, and some of the photosynthesized compound escaped.

Corals and their algal cells live together, apparently to the benefit of both—an example of the relationship known as symbiosis. We will find other examples in the next section.

1.7 **Energy from sulphide**

There is very little life on the bottoms of the oceans, at depths of several kilometres. It is far too dark for photosynthesis, too dark even for the human eye to detect any light. There is a sparse rain of dead and decaying organisms sinking from shallower depths, but no other foodstuffs derived from photosynthesis descend so deep. This sinking food is broken down by bacteria, which serve as food for echinoderms and molluscs, which in turn are eaten by fishes.

That is the general picture. In this otherwise almost empty habitat there are patches of most abundant life, supported not by photosynthesis but by another energy source. They are found around cracks in the ocean floor.

Asia and America are moving slowly apart, at a rate of a few centimetres per year, and this movement is opening up cracks in the floor of the Pacific. Molten rock from the earth's core fills the cracks, solidifies, and cracks again as the movement continues. At the cracks there are springs discharging warm (20°C) water, containing high concentrations of sulphide, into the cold (less than 5°C) sea.

A research submarine exploring these springs in 1977 found dense populations of large and colourful animals 2.5 km below the ocean surface (Fig. 1.10). These included large numbers of a tube-living worm up to one metre long (*Riftia pachyptila*, a member of the obscure phylum Vestimentifera) and clams 30 cm long (*Calyptogena magnifica*). Research soon showed that the luxuriant growth around the springs did not depend principally on their warmth, but on the sulphide.

Energy is released when sulphide (S^{2-}) is oxidized to sulphite (SO_3^{2-}) and eventually sulphate (SO_4^{2-}). Bacteria that depend on this energy source have long

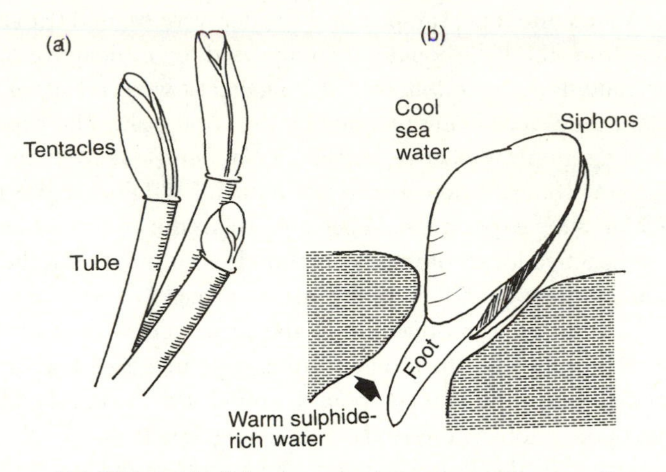

Fig. 1.10 Diagrams of (a) *Riftia* and (b) *Calyptogena* at a hot spring on the ocean floor.

been known from sulphide-rich terrestrial springs. They use energy from oxidizing sulphide in place of the solar energy used by photosynthesizing plants, to produce sugars and other foodstuffs from carbon dioxide.

Free-living sulphide-oxidizing bacteria are found in the deep-sea springs as well as the terrestrial ones. The key to the abundant animal life around the springs is that, as well as the free-living bacteria, there are symbiotic sulphide-oxidizing bacteria in the tissues of the worms and molluscs.

The giant worm *Riftia* (Fig. 1.10(a)) lives in white tubes with a plume of bright red tentacles protruding from the open end. It has no mouth and no gut and apparently cannot feed. Its largest internal organ, the trophosome, is a bag of sulphide-oxidizing bacteria. Sulphide and oxygen, and the carbon dioxide needed for synthesizing foodstuffs, diffuse from the water into the blood in the tentacles. The blood has a most unusual haemoglobin which binds both oxygen and sulphide and transports them to the trophosome. Human haemoglobin is made ineffective by sulphide which blocks the sites where oxygen would otherwise bind (that is one of the reasons why hydrogen sulphide is highly poisonous) but *Riftia* haemoglobin has separate binding sites for oxygen and sulphide.

The clam *Calyptogena* has its sulphide-oxidizing bacteria in its gills, where oxygen and carbon dioxide diffuse directly to them. It lives with its foot pointing down into one of the fissures from which the hot spring issues (Fig. 1.10(b)). The siphons, through which the water flowing over the gills enters and leaves, point upwards. Consequently, foot and gills are bathed by very different water. The foot is in the warm water flowing from the spring, rich in sulphide and almost devoid of oxygen. The gills have ordinary sea water flowing over them; it is cold and contains very little sulphide but some dissolved oxygen. Though the oxygen and carbon dioxide diffuse directly into the gills, the sulphide is taken up by the foot and transported to the gills in the blood. Unlike the giant worm's haemoglobin, that of the giant clam does not transport sulphide (and indeed can be poisoned by it). Instead sulphide is carried by a special large protein in the blood plasma. Bound to this protein, it does not affect the haemoglobin.

One of the problems of depending on energy from oxidization of sulphide is that access is needed to both oxygen and sulphide, but these react spontaneously if they meet. The sulphide comes from the hot spring water and the oxygen from the ordinary sea water, but if these waters mix the sulphide may be oxidized before it reaches the bacteria. The giant clam avoids the problem by using its foot to collect sulphide from the spring water while the oxygen diffuses into the gills from the sea water. The tube worm avoids it in a different way. In its environment, the emerging spring water and the sea water are not yet fully mixed, so the tentacles are exposed alternately to eddies of the two kinds of water and can capture sulphide and oxygen in turn. Once in the blood, they are kept apart by binding to different sites on the haemoglobin.

In this chapter we have seen that nearly all life depends on solar energy captured by photosynthesis; but nowhere is the sun's light used as fully as might seem theoretically possible. In the most productive terrestrial habitats, photosynthesis is limited by the sparse supply of carbon dioxide. In some others it is limited by water shortage. Primary production in plankton is limited by lack of nutrients. We have also seen how corals benefit from the photosynthesis of symbiotic algae and how animals at hot springs deep in the ocean have symbionts that are not dependent on solar energy, but on energy obtained by oxidizing sulphide.

2 Basic requirements

2.1 Energy from food

Having looked at the sources of energy for animal life, we now start to ask how much energy animals need. In this chapter we are concerned only with the energy used by resting animals; later chapters consider the energy needed for activities such as feeding and travelling from place to place, and also for growth. In this first section of the chapter we are concerned with methods for measuring the energy used by animals.

This energy is derived from food, so we are going to need to know how much energy is available from different kinds of food. Animals oxidize food to make the energy available, and we can find out how much energy is available by burning foodstuffs and measuring the heat that is produced. It does not matter that the food is oxidized in the animal by a series of enzyme-catalysed reactions, and in the experiment by burning; if the initial materials and the final products are the same, the same amount of energy will be released.

The heat produced when food is burned is usually measured by means of a bomb calorimeter (Fig. 2.1(a)). A weighed specimen of the food is put into a thick-walled metal container (the 'bomb') which is then filled with oxygen under pressure. The bomb is enclosed in an insulated container, then the specimen is ignited electrically. The heat produced warms the bomb and its temperature change is measured. In the version of the apparatus that is illustrated, a thermocouple is used to measure the temperature difference between the bomb and the metal block. If the heat capacity of the bomb is known, the heat can be

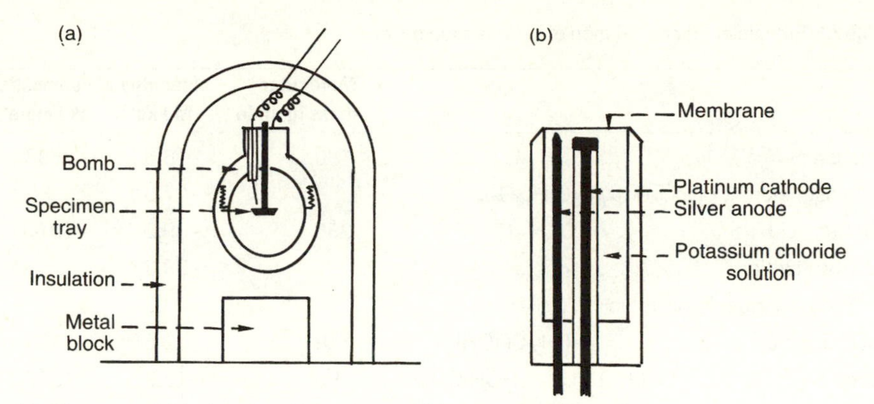

Fig. 2.1 Diagrams of (a) a bomb calorimeter and (b) an oxygen electrode.

calculated. The heat capacity is the energy needed to raise the temperature by one kelvin (one degree Celsius).

This heat measured by bomb calorimetry is not necessarily exactly equal to the quantity we want to know. There will be a discrepancy if burning the food causes a change in pressure in the bomb. A rise in temperature will of course tend to increase the pressure, but if the heat capacity of the bomb is large enough this effect will be small. The problem that concerns us arises in cases in which burning changes the number of gas molecules present. Consider, for example, combustion of palmitic acid (a fatty acid, $C_{16}H_{32}O_2$):

$$\underset{\text{solid}}{C_{16}H_{32}O_2} + \underset{\text{gas}}{23O_2} = \underset{\text{gas}}{16CO_2} + \underset{\text{liquid}}{16H_2O} \tag{2.1}$$

The number of molecules of gas falls and (since the volume of the bomb is fixed) the pressure falls. To restore the pressure to its initial value, the volume would have to be reduced; work would have to be done compressing the gases to this smaller volume, and this work would increase the energy of the system. The quantity of energy released by oxidizing palmitic acid at constant pressure (as in a living animal) is therefore a little greater than is released by oxidizing it at constant volume (as in a bomb calorimeter). The energy change at constant pressure is called the **enthalpy of combustion** and is the energy we want to know. Thus the result from bomb calorimetry has to be corrected to get a really accurate value for the enthalpy. However, the correction is very small, 0.2 per cent in the case we have been considering. In the case of carbohydrates, no correction is needed because the number of carbon dioxide molecules produced when they are oxidized equals the number of oxygen molecules used. Some examples of enthalpy of combustion are given in Table 2.1. Note that the enthalpy of combustion is the change in energy content of the food resulting from oxidation, so it is *minus* the heat production, and is negative in all these examples. The enthalpy of a particular food (for example, roast beef) can be determined directly by bomb calorimetry or estimated from the food's composition by using the data in the table.

Table 2.1 Enthalpies of combustion of various foodstuffs

		Molecular mass (g mole^{-1})	Enthalpy of combustion (MJ kg^{-1})	(kJ mole^{-1})
Glucose	$C_6H_{12}O_6$	180	–15.6	–2803
Cellulose, starch, glycogen	$(C_6H_{10}O_5)_n$		–17.5	
Palmitic acid	$C_{16}H_{32}O_2$	256	–39.2	–10 031
Fat from cattle			–39.3	
Mammalian muscle proteins			–23.7	
Succinic acid	$(CH_2{\cdot}COOH)_2$	118	–12.6	–1493
Lactic acid	$CH_3{\cdot}CHOH{\cdot}COOH$	90	–15.2	–1367
Ethanol	C_2H_5OH	46	–29.7	–1367
Methane	CH_4	16	–55.6	–890

Data are from K. L. Blaxter (1989) *Energy Metabolism in Animals and Man* (Cambridge University Press) and M. S. Kharasch (1929) *Journal of Research of the National Bureau of Standards* **2**, 596–627.

Not all the energy potentially available from an animal's food is actually used in its metabolism. Some is lost in faeces or urine, and some is incorporated in the body by growth. The energy M used in metabolism can be calculated if all the relevant enthalpies are known:

$$M = H_{\text{food}} - H_{\text{faeces}} - H_{\text{urine}} - H_{\text{growth}} \tag{2.2}$$

where the symbols H represent the enthalpies indicated by the subscripts.

2.2 Heat production

Another way of measuring the metabolic rate of an animal is to measure its heat output. Most of the energy released from food by an animal's metabolism immediately becomes heat, but some may be used to do work, and so converted to other forms of energy. If I climb a mountain, the potential energy (energy due to height) of my body increases, and as I descend my potential energy falls. In the ascent, some of the energy from my metabolism becomes potential energy, but in the descent this potential energy becomes heat. If I start running, the kinetic energy (energy due to motion) of my body increases, but when I decelerate at the end of a race my kinetic energy falls again. As I accelerate, metabolic energy is converted to kinetic energy but when I decelerate this energy is returned as heat. Suppose that an animal is in the same state at the beginning and end of a period of measurement (for example, resting at sea level). Suppose also that it has not done work on its environment, for example by lifting weights on to a shelf. In that case, the energy released by its metabolism has all become heat, and its metabolism can be measured by determining its heat output.

This is done in a chamber called a calorimeter. Two designs are shown in Fig. 2.2. Design (a) is heavily insulated and would get gradually hotter as the animal

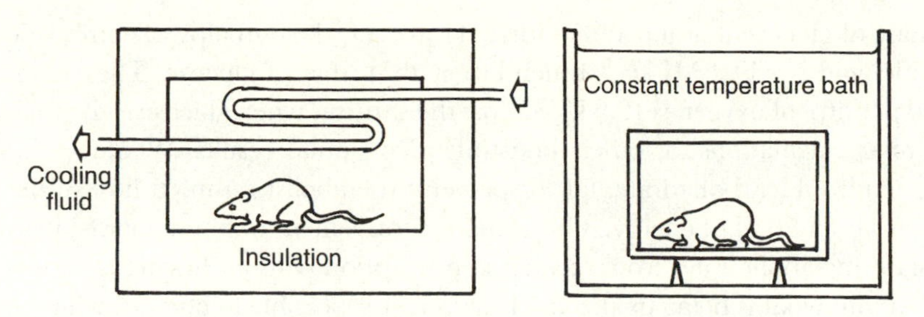

Fig. 2.2 Two designs of calorimeter used for measuring metabolic rates.

metabolized if it were not for a cooling system that keeps its temperature constant; cool fluid flowing through a heat exchanger removes the heat produced by metabolism. If the rate of flow of fluid and its change of temperature are known, the rate of heat production can be calculated.

An alternative design (Fig. 2.2(b)) has thinner insulation, thin enough to ensure that even without a heat exchanger the temperature rises only a little. In the example shown, the calorimeter is immersed in a constant temperature bath. The animal is left in the calorimeter for long enough for it to reach equilibrium at a steady temperature (assuming that the animal is producing heat at a constant rate). Then the temperature difference across the insulated wall of the calorimeter is measured. From this, the rate of metabolic heat production can be calculated, using the standard equation for thermal conduction (eqn 6.2).

2.3 Respiratory gases

Respiration uses up oxygen and produces carbon dioxide, so these gases can be used to determine metabolic rates. For example, consider the oxidation of glucose, $C_6H_{12}O_6$:

$$C_6H_{12}O_6 + 6O_2 = 6CO_2 + 6H_2O \tag{2.3}$$

The molecular mass of glucose is 180 g mol^{-1}, so one mole of glucose is 180 grams. One mole of a gas occupies 22.4 litres at 0°C and one atmosphere pressure, so six moles either of oxygen or of carbon dioxide occupy $6 \times 22.4 = 134.4$ litres. Equation 2.3 tells us that six moles or 134.4 litres of oxygen are needed to oxidize one mole or 180 grams of glucose. One litre of oxygen oxidizes $180/134.4 = 1.34$ grams of glucose, and one litre of carbon dioxide is produced in this oxidation. The enthalpy of combustion of glucose is -15.6 MJ kg^{-1} (Table 2.1) so oxidation of 1.34 g (0.00134 kg) releases 0.0209 MJ (20.9 kJ). If one litre of oxygen is used to oxidize glucose, 20.9 kJ of energy are made available.

That conclusion applies specifically to glucose. Now consider the case of palmitic acid. Equation 2.1 shows one mole of this fatty acid, with a mass (calculated as we did for glucose) of 256 grams, reacting with 23 moles of oxygen, with a volume of 515 litres. A litre of oxygen oxidizes only 0.50 grams of palmitic acid, far less than the mass of glucose that it could oxidize. However, the enthalpy of combustion of palmitic acid is –39.2 MJ kg^{-1}, much larger than that of glucose. The energy released per litre of oxygen is 19.6 kJ, almost the same as when glucose is the fuel.

Similar calculations for other foodstuffs give similar results. Whatever food is being oxidized (carbohydrate, fat, or protein) metabolism using a litre of oxygen releases about 20 kJ of energy. This is highly convenient to physiologists, who can calculate metabolic rates from oxygen consumption without having to worry too much about what is being oxidized. However, it is possible to find out what classes of foodstuff are being oxidized if further measurements are taken. If carbon dioxide production is measured as well as oxygen consumption, carbohydrate metabolism can be distinguished from fat metabolism. When carbohydrates are being oxidized, the volume of carbon dioxide produced equals the volume of oxygen used (notice that eqn 2.3 has $6O_2$ on the left-hand side and $6CO_2$ on the right). When fat is being oxidized, only about 0.7 volumes of carbon dioxide are produced for each volume of oxygen (eqn 2.1 shows a ratio of 16/23 for a fatty acid). Further, when protein is oxidized, nitrogenous waste (usually ammonia or urea) is produced. If oxygen, carbon dioxide, and nitrogenous wastes are all measured, the proportions of carbohydrate, fat, and protein being metabolized can be estimated and the energy released can be calculated more accurately.

Such precision is often not needed. It is generally sufficient to measure only oxygen. Aquatic animals such as fish may be kept in closed containers of water and the concentration of dissolved oxygen measured from time to time. Terrestrial animals may be kept in a closed container of air and the concentration of oxygen in that measured periodically. In neither case must the concentration be allowed to fall far, or the animal may suffer; its metabolism will be affected and it will eventually suffocate.

Oxygen electrodes (Fig. 2.1(b)) are very useful for measuring oxygen concentrations in water. An oxygen electrode actually contains two electrodes, a silver anode and a platinum cathode, with a small potential difference between them. They are surrounded by a potassium chloride solution which is separated by a plastic membrane from the fluid whose oxygen content is to be measured. Any oxygen that diffuses through the membrane into the potassium chloride is immediately reduced by an electrochemical reaction, and an electric current flows. Thus the partial pressure of oxygen in the potassium chloride solution remains close to zero; the rate of diffusion of oxygen through the membrane is proportional to its partial pressure in the fluid being tested and the current through the electrode is also proportional to the partial pressure of oxygen.

2.4 Minimal metabolism

My car consumes no fuel when I park it and switch off the engine, but the processes of life continue to use energy even when an animal is resting. This energy consumption is the subject of this section.

Unfortunately, it is hard to define the conditions in which it should be measured. An animal's metabolic rate rises when it moves or feeds. It remains high for a while after a meal while digestion continues. It is affected by temperature. In humans it rises when we get excited and falls when we go to sleep.

The concept of **basal metabolism** has been defined carefully for humans. The subject should be resting, lying but awake. She or he should not have starved too long or eaten too recently; measurements are often made 10–12 hours after a moderate evening meal. Emotional stress should be avoided, and the subject should be familiar with the surroundings. The temperature should be comfortable, in the thermoneutral zone (defined in Chapter 6).

Careful planning is needed, even with human subjects, to get all these conditions right, and they generally cannot be fully achieved for animals; try telling a worm to stop wriggling or a cockroach to lie still. The term **standard metabolism** is often used when the conditions are carefully defined but not necessarily exactly the same as for basal metabolism. **Resting metabolism** means (obviously) that the animal was resting, but may have eaten recently. **Fasting metabolism** means that it has not eaten for some time but may move. A strategy that can be useful with animals that cannot be prevented from moving is to make a continuous record of metabolic rate, which will increase as the animal starts moving and stop when it stops. The lowest observed metabolic rate will presumably correspond to a period of rest. We will use the term **minimal metabolism** to apply indiscriminately to any of these definitions.

Endotherms ('warm-blooded' animals) have a thermoneutral range of temperatures over which standard metabolism is constant and has its lowest value. The metabolic rates of ectotherms ('cold-blooded' animals) are lowest when the animals are cold and rise steadily as their temperature increases, so measurements have little value unless the temperature is stated.

Humans have basal metabolic rates around 80 watts, so a resting human emits a little less heat than a 100 watt light bulb. Larger mammals have higher metabolic rates and smaller ones have lower ones, but the rates are not proportional to body mass. For example, a two tonne elephant is 100 000 times as heavy as a 20 gram mouse but has only 5000 times the metabolic rate (2000 watts compared with 0.4 watts). Weight for weight, it is a great deal cheaper to feed elephants than mice.

Two studies published independently in 1932, one by Kleiber and the other by Brody and Procter, showed a clear relationship between metabolic rate and body mass for mammals. An ordinary graph of metabolic rate against mass would be a curve, but when the graph is plotted on logarithmic coordinates as in Fig. 2.3(a) all the points lie close to a straight line. Notice that in this graph, factors of 10 are spaced equally along the axes; 100 kg is as far from 10 kg as 10 kg is from 1 kg. Remember that the logarithms to base 10 of 1, 10, and 100 are 0, 1, and 2 respectively. In effect, Fig. 2.3(a) is a graph of the logarithm of minimal metabolic rate against the logarithm of body mass. The straight line has a gradient of 0.76 and its equation is

$$\log_{10}R = 0.76\log_{10}m + 0.52 \tag{2.4}$$

where R is the minimal metabolic rate in watts and m is body mass in kilograms. By taking antilogarithms this equation becomes

$$R = 3.3m^{0.76} \tag{2.5}$$

Earlier writers had claimed that the metabolic rate was proportional to the surface area of the body. If animals of different sizes were geometrically similar, their areas would be proportional to (body length)^2 and their volumes to (body length)^3, so surface area would be proportional to (volume)$^{2/3}$. If different-sized animals are made of similar materials, with the same density, that implies that surface area should be proportional to (body mass)$^{2/3}$. Different-sized animals are not geometrically similar (a mouse is not the same shape as a shrunk elephant) but the prediction is not far wrong; surface areas of birds and mammals over wide

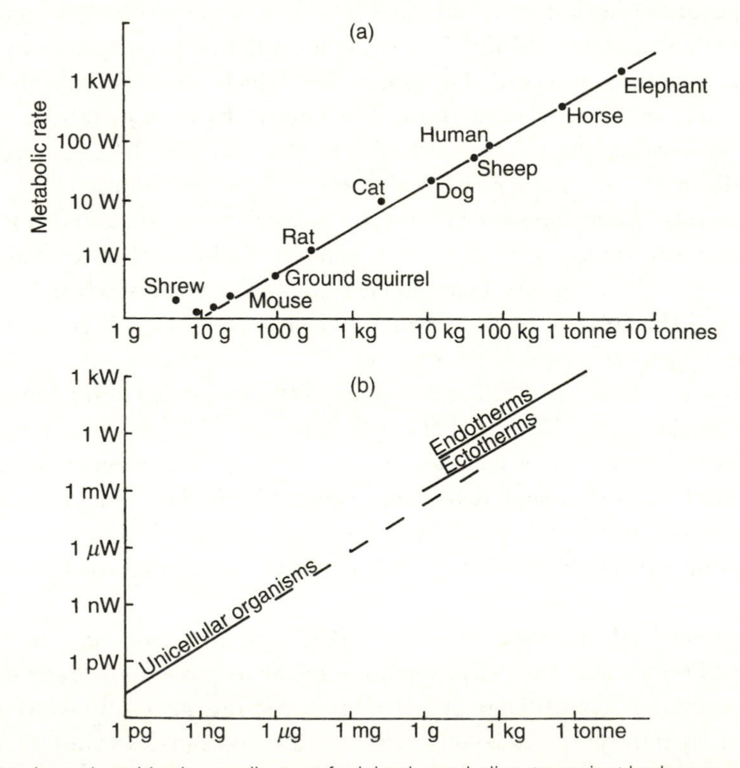

Fig. 2.3 Graphs on logarithmic coordinates of minimal metabolic rate against body mass. (a) Data for mammals alone. The line represents eqn 2.5. (b) Data for endotherms (mammals and birds), multicellular ectotherms, and unicellular organisms.(The data are from tables in K. Schmidt-Nielsen (1983) *Animal Physiology: Adaptation and Environment*, Cambridge University Press and R. H. Peters (1983) *The Ecological Implications of Body Size*, Cambridge University Press.)

ranges of size have been measured and found to be proportional to (body mass)$^{0.67}$ and (body mass)$^{0.65}$ respectively. However, statistical analysis of the data in Fig. 2.3(a) shows that the gradient is significantly greater than 0.67. Minimal metabolic rate is not proportional to the surface area of the body.

Figure 2.3(b) shows minimal metabolic rates for other animal groups as well as mammals. The line for each major group has about the same gradient as the mammal line, but the lines are at different levels. The (endothermic or 'warm blooded') birds and mammals use energy much faster than equal-sized (ectothermic or 'cold blooded') reptiles and fishes. Birds and mammals generally have body temperatures between 36 and 43°C, and the metabolic rates of ectotherms shown in Fig. 2.3(b) were measured at only 20°C, but this difference is not enough to explain the difference in metabolic rate. The preferred body temperatures of many reptiles are similar to those of mammals (Chapter 6 explains how they warm themselves by basking in the sun), and metabolic rates of some of them have been measured in this temperature range. The minimal metabolic rates of lizards at 37°C are about five times the rates at 20°C, but still only about one fifth of the rates for similar-sized birds and mammals.

Figure 2.3(b) also shows that the metabolic rates of unicellular organisms are lower than those of multicellular ectotherms of equal mass. For example, a large protozoan has only about one quarter the metabolic rate of a copepod of equal mass.

Table 2.2 gives equations relating minimal metabolic rate to body mass, for various groups of animals. The exponents vary a little, but all of them are fairly close to 0.75. It seems to be a very general rule that the minimal metabolic rates of related animals are about proportional to (body mass)$^{3/4}$. The most successful

Table 2.2 Equations* showing how minimal metabolic rate (in watts) is related to body mass (in kilograms) for various groups of animals

	Body temperature	**Factor, *a***	**Exponent, *b***
Mammals	Normal	3.3	0.76
Passerine birds	Normal	6.3	0.72
Other birds	Normal	3.6	0.72
Lizards	37°C	0.68	0.82
Lizards	20°C	0.13	0.80
Fishes	20°C	0.43	0.81
Crustaceans	20°C	0.27	0.78
Unicellular organisms	20°C	0.055	0.83

*The equations have the form

metabolic rate = a(body mass)^b

where a and b are constants.

The data are from various sources, tabulated by R. H. Peters (1983) *The Ecological Implications of Body Size*, Cambridge University Press.

attempt to explain this up to now was published in 1997. It is a theory that can apply to any group of animals or plants in which oxygen or other resources are distributed by a branching system of tubes: for example by an arterial system or, as in insects, by a branching system of tracheae.

The theory assumes that the distributing system is fractal-like, which means that if a small part of the system is magnified it will look just like the whole system. Figure 2.4(a) shows the general idea, but whereas all the finest branches are to the right of the diagram, the animal would have fine branches all through its volume. In a true fractal the branching would continue indefinitely, and the finest branches would be infinitesimally small. The theory postulates, however, that the finest branches are capillaries whose dimensions are the same in animals of all sizes. This is a reasonable assumption; for example, blood capillaries are about the same size in mammals ranging from shrews to elephants.

From this point I will use the language of blood systems, referring to the large branches as arteries and the smallest ones as capillaries; but please remember that the argument applies equally to tracheal systems. Figure 2.4(a) shows the blood vessels dividing into three at each branching point, but from this point I will assume instead that they divide into eight. The argument leads to the same conclusion, however many branches there are at each branching point, but the choice of eight simplifies the arithmetic that follows.

In Fig. 2.4(b), a cube is shown divided into eight smaller cubes, each of which is divided into eight still smaller cubes. Imagine this cube as part of an animal, with each of the smallest cubes served by a single capillary. Each block of eight of the smallest cubes is served by an arteriole which divides to form their eight capillaries. This block has twice the side length of the smallest cubes, so the arteriole must be twice as long as the capillaries (Fig. 2.4(c)). It carries the blood to

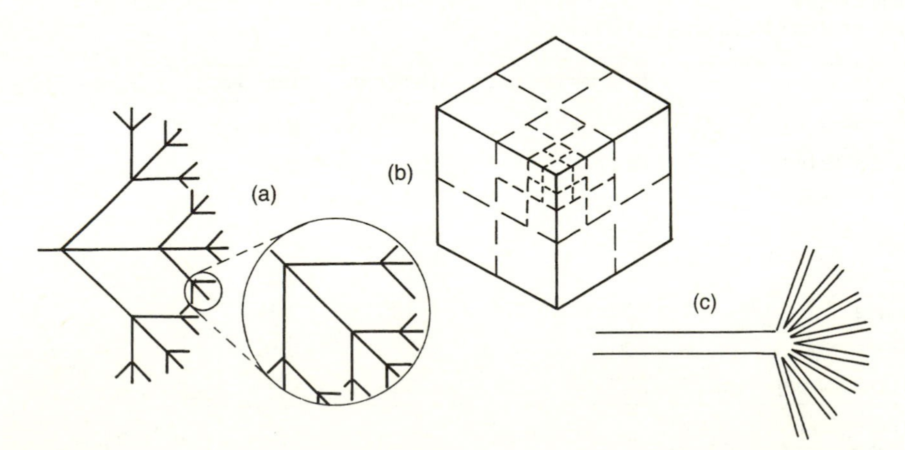

Fig. 2.4 Diagrams illustrating a theory relating metabolic rate to body mass. (a) A diagram of a fractal branching system. A small part, enlarged, resembles the whole system. (b) A cube divided into eight smaller cubes, which in turn are subdivided. (c) A single branch point, as envisaged in the theory.

all eight capillaries, so needs to be wider than they are; the theory assumes that its cross-sectional area is eight times that of the capillaries. Because volume is length times cross-sectional area, the volume of the arteriole is 16 times that of the capillary. More generally, it is assumed that at every level of branching, a vessel divides into eight branches, each of half its length, one eighth its cross-sectional area, and one sixteenth its volume.

Our aim is to explain why minimal metabolic rates of similar animals tend to be proportional to (body mass)$^{3/4}$. We will tackle the problem by assuming that metabolic rate is proportional to (body mass)^a, where a is some constant. Can we predict the value of a?

Suppose that a capillary can supply enough oxygen to satisfy the needs of an animal of mass m_0. Then an animal of mass m has $(m/m_0)^a$ times the metabolic rate of one of mass m_0 and needs $(m/m_0)^a$ capillaries. If an animal's blood vessels divide once it has eight capillaries, if there are two levels of branching it has $8^2 = 64$ capillaries, and if there are N levels of branching it has 8^N capillaries. Our animal of mass m needs N levels of branching, where

$$8^N = (m/m_0)^a \tag{2.6}$$

Now assume that the blood systems of different-sized animals occupy equal fractions of their body volumes. An animal with 8^N capillaries each of volume V_{cap} has a total volume $8^N V_{cap}$ of capillaries. It has 8^{N-1} arterioles each of volume $16V_{cap}$, with a total volume of $8^{N-1}16V_{cap} = 2 \times 8^N V_{cap}$. It has a total volume $4 \times 8^N V_{cap}$ of vessels of the next size, and so on until, finally, it has eight of the largest size of arteries with a total volume $2^{N-1} \times 8^N V_{cap}$. The total volume of the blood system is

$$V_{total} = [1 + 2 + 4 + 8 + \ldots 2^{N-1}] \times 8^N V_{cap}$$

When $N = 1$, the term in the square brackets is 1, which equals $(2 - 1)$; when $N = 2$ it is $[1 + 2]$, equal to $(4 - 1)$; when $N = 3$ it is $[1 + 2 + 4]$, equal to $(8 - 1)$; when $N = 4$ it is $[1 + 2 + 4 + 8]$, equal to $(16 - 1)$; and so on. Thus, if N is reasonably large (as it will be, in all but the smallest animals), the term in square brackets is approximately equal to 2^N and

$$V_{total} \approx 2^N \times 8^N V_{cap} = 16^N V_{cap} \tag{2.7}$$

If the blood system occupies the same fraction of the volume of the body in animals of all sizes, $m/m_0 = V_{total}/V_{cap}$, and eqns 2.6 and 2.7 give us

$$8^N = (16^N)^a = 16^{Na}$$
$$2^{3N} = 2^{4Na}$$
$$a = 3/4$$

Thus the theory predicts what has been found to be the case, that the metabolic rates of similar animals will be proportional to (body mass)$^{3/4}$.

Some questions remain, about how soundly the theory is based and how widely it applies. Clearly, it applies to vertebrates and to invertebrates such as squids in which the arteries divide to form, eventually, capillaries which permeate the tissues. It also applies to insects, in which oxygen is distributed to the tissues by the tracheal system. But it is not at all obvious how it can apply to animals such as crustaceans and snails, which have open blood systems, with major arteries discharging into the body cavity; or to unicellular organisms and lower invertebrates which have no blood system. Yet these organisms also have metabolic rates roughly proportional to (body mass)$^{0.75}$ (Table 2.2).

Even the applications to vertebrates and insects have problems. We assumed that total cross-sectional area remained constant, for successive levels of branching (a major branch dividing into eight branches, each of one eighth the cross-sectional area). For vertebrates, that would imply that blood would flow at the same speed in the capillaries as in the aorta. In reality, the smallest vessels are wider than the theory assumes, and blood flows much more slowly in the capillaries than in major arteries, which it needs to do to allow time for gas exchange.

In insects, air may be pumped through the larger tracheae, but oxygen moves through the finer branches solely by diffusion. The total area of branches at any level is assumed to be proportional to the metabolic rate, but larger animals have more levels, so the diffusion distances in them are longer. This implies problems for larger insects, in getting adequate supplies of oxygen to their tissues.

Another problem with the theory is that the capillaries are assumed to be the same length in animals of all sizes, although each capillary has to supply a bigger block of tissue in larger animals, which have lower metabolic rates per unit volume. Yet another possible criticism is that the theory makes the path through the blood vessels—from the heart to the capillaries—the same length for all parts of the body. In real animals, this path is shorter for parts of the body that are close to the heart.

2.5 Metabolic rate and lifestyle

Equations like these do not tell the whole story. There are differences of metabolic rate between related animals of similar size which seem to be related to differences in their habits or environments. Figure 2.5 shows this for mammals. In Fig. 2.3, minimal metabolic rate was plotted on logarithmic coordinates against body mass and lines with gradients around 0.75 were obtained. In contrast, in Fig. 2.5 (minimal metabolic rate/body mass) is plotted against body mass and the gradients of the bands are about –0.25. This graph shows that in general, mammals that eat grass, nuts, or other vertebrates have high metabolic rates for their masses, in contrast to those that eat invertebrates, fruit, and the leaves of trees and have low metabolic rates for their masses. Some specific examples (all mammals of 3–4 kg) are given in Table 2.3. Among these particular examples, the mangabey (a fruit-eating monkey) has a metabolic rate close to the

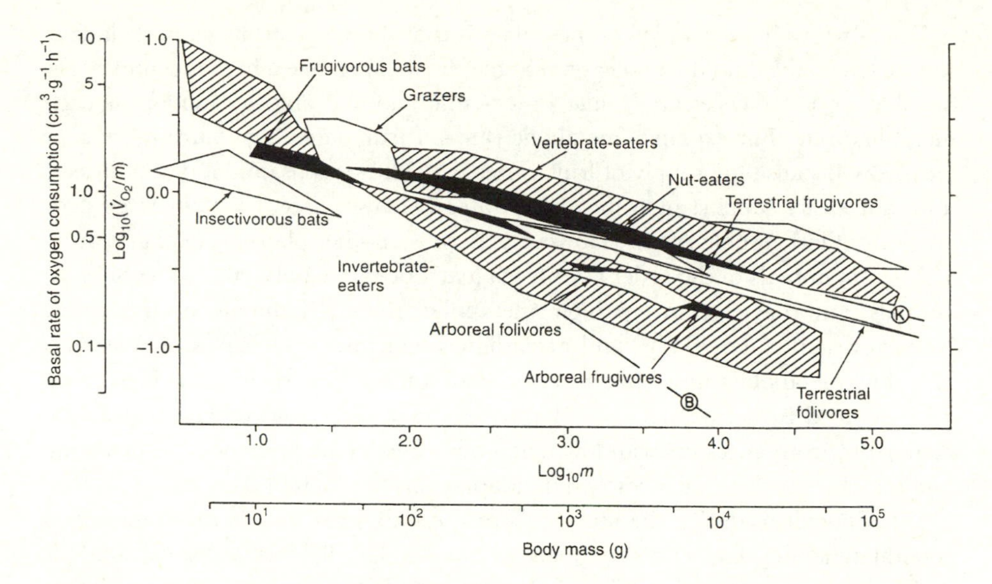

Fig. 2.5 Minimal metabolic rates of mammals that eat different diets. Metabolic rate per unit body mass is plotted against body mass, on logarithmic coordinates. Oxygen consumption at a rate of 1 $cm^3\ g^{-1}\ h^{-1}$ releases energy at a rate of 5.6 $W\ kg^{-1}$. Each polygon encloses all the points for animals with a particular feeding habit. Line K represents equation 2.5 (After B. K. McNab (1986) *Ecological Monographs* **56**, 1–19.)

Table 2.3 Minimal metabolic rates of similar-sized mammals that eat different diets

Species	Mass (kg)	Food	Metabolic rate ($W\ kg^{-1}$)
Wildcat, *Felis sylvestris*	3.3	Vertebrates	3.9
Coypu, *Myocastor coypus*	4.3	Grass	3.9
Jack rabbit, *Lepus alleni*	3.4	Grass	2.5
Agouti, *Dasyprocta azarae*	3.8	Seeds and nuts	2.7
Mangabey, *Cercocebus torquatus*	4.1	Fruit	2.3
Anteater, *Tamandua tetradactyla*	3.5	Insects	1.4
Armadillo, *Cabassous centralis*	3.8	Insects	1.2
Sloth, *Bradypus varigatus*	3.8	Leaves of trees	1.0

Data are from M. A. Elgar and P. H. Harvey (1987) *Functional Ecology* **1**, 25–36.

grass-eating jack rabbit, but otherwise they illustrate the general conclusion that comes from Fig. 2.5: mammals that eat grass, nuts, or vertebrates have higher minimal metabolic rates than those that eat insects or leaves from trees. I would like to emphasize that we are concerned here with *minimal* metabolic rates. Insectivorous bats have high metabolic rates when they are flying, but their minimal metabolic rates are low, in comparison with other mammals of equal mass.

The obvious interpretation of these data is that there is a strong causal relationship between diet and metabolic rate. Some diets may make a high metabolic rate possible or even necessary, while others may not be able to support a high metabolic rate. For example, metabolic rates of fruit-eaters may have to be kept fairly low because the supply of fruit varies with the seasons, and a diet of insects may not easily support high metabolic rates because insects contain large proportions of indigestible cuticle. However, there is another plausible explanation.

Most mammals of the order Carnivora (cats, dogs, weasels, etc.) eat vertebrate animals and have relatively high metabolic rates. Mammals of the order Xenarthra (anteaters, sloths, and armadilloes) eat insects or leaves of trees and have low metabolic rates. May it not be that some orders such as the Carnivora have evolved from ancestors with high metabolic rates, and others such as the Xenarthra from ancestors with low metabolic rates? The differences of metabolic rate may be accidents of ancestry, not adaptations to different diets.

It is difficult to decide the answer from the data available, because there is a general tendency for related animals to eat similar diets. A statistical analysis designed to answer such questions led to the conclusion that the only association between metabolic rate and diet that could not be explained by ancestry was the one between a diet of vertebrates and high metabolic rate. The analysis does not tell us that metabolic rates of mammals depend more on ancestry than on diet, but merely that the opposite cannot be proved.

Fishes swimming deep in the oceans seem to have lower metabolic rates than those that swim nearer the surface. This conclusion came from a study of teleost fish that swim in mid water (well clear of the bottom) off the Californian coast. Each species was caught over a range of depths, in some cases a very large range because many of them spend the night much nearer the surface than the day. The depth used in the analysis was the 'minimum depth of occurrence', defined as the depth below which 90 per cent of catches were made. Figure 2.6(a) shows minimal metabolic rate plotted against this minimum depth. Temperature was higher near the surface, but the metabolic rates were measured at 5°C. There seems to be a clear tendency for metabolic rate to be less for deeper-living fish, but there are several points that we should consider.

First, the fish varied in size and, unfortunately for the analysis, all the larger species lived relatively deep. The individual fishes had masses ranging from about 1 to 100 grams, with quite a large range of sizes for some species. All species that included individuals of 20 grams or more had minimum depths of occurrence of 300 metres or more. This may have exaggerated the apparent effect of depth because metabolic rate/body mass is generally lower for larger species. However, small (<5 g) species occur throughout the range of depths and support the conclusion.

Next, we should notice that although Fig. 2.6 includes data from fishes belonging to 12 different families, all the species going within 200 metres of the surface belong to just two families, the Bathylagidae and the Myctophidae. Could it be

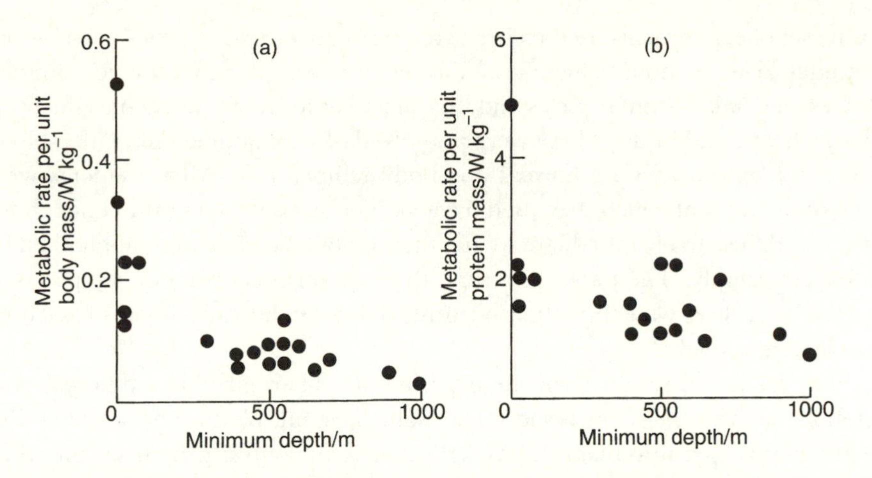

Fig. 2.6 Metabolic rates of mid-water fishes. (a) Metabolic rate/body mass; (b) metabolic rate/protein mass; both plotted against the minimum depth of occurrence. (Data from J. J. Torres *et al.* (1979) *Deep-Sea Research* **26A**, 185–97.)

that metabolic rate depends more on ancestry than depth? Conveniently, it seems possible to discount this possibility because both the Bathylagidae and the Myctophidae include deep-living species with low metabolic rates as well as shallower-living species with higher rates. However, this has not been tested formally using statistics.

The fish that swim deep have very watery flesh. Could their low metabolic rates simply be the effect of dilution of the tissues with non-metabolizing water? To check this possibility the data have been re-plotted in Fig. 2.6(b) with metabolic rate divided by the mass of protein in the body instead of by body mass. One surface-living species has a high metabolic rate even by this criterion, but the other surface-living species have metabolic rates per unit protein close to those of species from a few hundred metres deeper. The low metabolic rates of the deep-living species may be at least partly due to their tissues being diluted with water.

2.6 How energy is used in minimal metabolism

Now that we know how much energy is used in minimal metabolism, we ask, what is it used for? An obvious first step is to find out which parts of the body are using energy fastest, in a resting animal.

Tissues taken from a freshly killed or anaesthetized animal do not immediately die, but will continue their metabolism for hours if kept in suitable conditions. They must be kept in a solution that resembles body fluids. And, because the heart is no longer pumping blood through their blood vessels, something must be done to ensure that they get the oxygen that they need for their metabolism.

One set of experiments used tissues taken from anaesthetized dogs (which were eventually killed without being allowed to recover from the anaesthetic). Bundles of fibres were taken from muscles and thin slices cut from organs such as the liver, kidneys, heart, and brain. These were put into flasks containing a little dog blood serum and kept at 38°C, a normal dog body temperature. Measurements were made of the rates at which they used oxygen. The slices were cut thin enough for oxygen to diffuse freely into them, so that their metabolic rates were not limited by the oxygen supply. The paper reporting the experiments does not say precisely how thick the slices were, but other scientists doing similar experiments have used slices 0.25 mm thick.

Table 2.4 shows results from the experiments. There are huge differences in metabolic activity, between tissues. The skin, skeleton, blood, and fat have low metabolic rates per unit mass, 1.4 W kg^{-1} or less. In contrast, the liver, intestinal mucosa, kidneys, and brain have high rates, 7.6 W kg^{-1} or more. (The mucosa is the inner lining of the gut wall.) The low rates do not seem surprising for the particular tissues concerned, but the very high rates of 13.7 W kg^{-1} for kidney and 11.4 W kg^{-1} for liver are striking.

The next column in the table shows the total mass of each tissue in the body, determined by dissecting the dogs at the end of the experiment. The third column of figures shows the mass of each tissue multiplied by its metabolic rate per unit

Table 2.4. Metabolic rates of organs of 20 kg dogs, calculated from the results of experiments with small pieces of tissue

Tissue	Metabolic rate/mass (W kg^{-1})	Mass (kg)	Metabolic rate of organ (W)
Muscle	3.2	8.14	25.8
Skeleton*	0.2	4.26	0.7
Skin	0.9	2.57	2.4
Blood	0.03	1.33	0.04
Fat	1.4	0.78	1.1
Liver	11.4	0.44	5.0
Small intestine mucosa	7.8	0.24	1.9
Stomach mucosa	5.0	0.10	0.5
Heart	4.2	0.15	0.6
Kidneys	13.7	0.09	1.2
Brain	7.6	0.11	0.9
Remainder		1.89	2.1
Total		20	42.2

Data are from A. W. Martin and F. A. Fuhrman (1955) *Physiological Zoology* **28**, 18–34.
*Skeleton mass is larger than the mass of bone because the bones were cleaned only roughly, leaving fragments of muscle attached and the marrow still inside. Also, there is some doubt about skeletal mass because there is a discrepancy between the text and a table in Martin and Fuhrman's paper.

mass (the first column multiplied by the second) to get an estimate of the contribution each tissue makes to the metabolic rate of the complete animal. It shows muscle making much the largest contribution, with the liver next in importance.

We must ask whether the tissue slices in the experiment metabolized as fast as they did in the intact animal. It appears that, on average, they did. The total metabolic rate obtained by adding together all the estimates for individual organs is 42.2 watts. Metabolic rates of intact anaesthetized dogs of the same body mass as those from which the slices were taken were found to be almost the same, 40 watts. Such close agreement must be to some extent coincidental, but if some tissues metabolized faster in the intact animal, others metabolized less fast.

Similar experiments have been done with tissue slices from rats and mice. They gave essentially similar results, showing that the kidneys, liver, and brain had the highest metabolic rates per unit mass and that muscle and liver made the biggest contributions to whole-body metabolic rate. The total metabolic rates calculated from the experiments with individual tissues were 66 per cent (rat) and 72 per cent (mouse) of the minimal metabolic rates of the intact animals.

The agreement between the results from tissue slices and from intact animals may seem less good for rats and mice than for dogs, but in this respect the rat and mouse results are perhaps more believable than the dog ones. We should expect the totals from the tissue-slice experiments to be lower than the metabolic rates of the intact animals, because the animals were doing things that the slices could not do. Most obviously, they were breathing and their hearts were beating.

The heart of a resting 20 kg dog pumps about 50 cm^3 (5×10^{-5} m^3) blood per second. Most of this leaves the left ventricle at a pressure of 15 000 Pa. The power exerted by a pump is the flow rate multiplied by the pressure difference across the pump, in this case $5 \times 10^{-5} \times 15\,000 = 0.75$ watts. Muscle generally works with efficiencies around 20 per cent, as will appear in Chapter 4, so for the heart to deliver 0.75 watts of mechanical power it must use about 4 watts of metabolic power. To that estimate for the left side of the heart we should add a little for the right side (which pumps the same volume of blood at much lower pressure, so uses less power): for the whole heart we will estimate 5 watts. This is about 13 per cent of the dog's metabolic rate. A similar calculation indicates that the cost of breathing is much less than the cost of circulating the blood; larger volumes of air than of blood are pumped, but the pressure is much less.

Another way to find out how much energy each organ uses is to measure the flow of blood through the organ in an intact animal, and the oxygen content of the blood arriving in the organ's artery and leaving in its vein. The rate at which oxygen is being used is (blood flow rate) × (change in oxygen concentration). Such experiments involve surgical operations which some biologists prefer not to perform, but which need cause the animal no apparent distress if done skilfully and humanely. They are subject to strict legal controls, of which the details vary between nations.

In one set of experiments, oxygen consumption of the gut of dogs was investigated. A cannula was implanted in the aorta of an anaesthetized dog so that small samples of blood could be withdrawn for analysis. The aorta supplies blood to all parts of the body. Another cannula was put in the portal vein, which carries the blood leaving the gut, taking it to the liver. Also, an electromagnetic flow meter was fitted to the portal vein; this instrument makes it possible to measure the rate of flow of blood without having to cut the vein open. This experiment showed that the blood in the portal vein contains 4 cm^3 less oxygen, per 100 cm^3, than aortic blood. The flow rate in the portal vein was 10 $cm^3\ s^{-1}$ (this is the rate for 20 kilogram dogs, estimated from the rate that was actually measured in smaller dogs). Thus the gut was using $4 \times 10/100 = 0.4\ cm^3$ oxygen per second. Metabolism using 1 cm^3 oxygen releases 20 joules, so the gut's metabolic rate must have been 8 watts, 20 per cent of the expected metabolic rate for a 20 kilogram dog. The rate would have varied considerably, depending on how recently the dog had eaten.

We now know about how much metabolism is happening in each of the principal organs of a resting dog, and also that only a small proportion of the body's metabolism is needed to drive breathing and the heartbeat. How is the rest of the energy being used?

Table 2.4 tells us that metabolism per unit mass is fastest in the kidneys. This is suggestive. The mechanism of the mammalian kidney, producing concentrated urine, depends on sodium ions being moved by active transport through the walls of the kidney tubules. Active transport of sodium ions occurs especially rapidly in the kidneys, but also occurs in every other tissue. This is because there is a marked difference in ionic composition between cell contents and extracellular body fluids. Body fluids including blood plasma have a relatively high sodium concentration and a low potassium concentration. Cell contents have the reverse. Consequently, sodium is constantly diffusing into cells and potassium is diffusing out. The sodium/potassium exchange pump in the cell membrane maintains the concentration difference by pumping sodium out of cells, and simultaneously pumping potassium in. Metabolic energy is needed to drive the pump. It has to work fast in the brain as well as the kidneys, because the action potentials used to transmit messages around the nervous system depend on the nerve cell membrane being made very briefly more permeable to ions. Every time an action potential passes, a little extra sodium is allowed to enter the cell and a little potassium is allowed out. Every action potential makes a little extra work for the sodium/potassium pump. There is a drug called ouabain that inhibits this pump but has no other known effect on cell function. Metabolic rates of rat liver, kidney, and brain slices have been measured, in solutions that did, or did not, contain ouabain. Metabolic rates in the presence of the drug were only about 45 per cent of normal values for kidney and brain, and 55 per cent for liver. If ion pumping is indeed the only process affected by ouabain, about half the metabolic rate of each tissue was being used to pump ions. Similar experiments on muscle indicate that only 5–10 per cent of its resting metabolism is powering the sodium/potassium pump, but we have to consider another ion pump as well. Muscles maintain a marked difference in calcium concentration between the cytoplasm (low concentration) and the contents of the sarcoplasmic reticulum (high concentration). This is necessary because the trigger for muscle contraction is rapid release of calcium ions from the reticulum into the cytoplasm. It has been estimated that calcium pumping, to maintain the concentration difference, accounts for 7 per cent of muscle's resting metabolic rate. Thus the total cost of ion pumping in muscle probably accounts for about 15 per cent of its metabolic rate. Taking account of the different costs of ion pumping in different tissues, we can estimate very roughly that about 9 watts or 23 per cent of the minimal metabolic rate of a 20 kilogram dog is used for ion pumping.

Another important energy-demanding process is protein synthesis. Even in animals that are not growing, protein is constantly being built and broken down. The rate can be measured by giving animals food containing radioactive amino acids. In some experiments, [^{13}C]leucine has been fed, and the blood serum analysed for the isotope. In others, [^{15}N]glycine has been fed and the isotope content of the urine measured. In either case, granted certain assumptions, the rate of protein turnover can be calculated.

In one set of experiments, dogs were found to turn over daily 12 grams of protein per kilogram of body mass, 240 grams per day for a 20 kilogram dog. I cannot explain why turnover happens at this particular rate, neither more nor less. Proteins consist of chains of amino acids connected by peptide bonds, and it is these bonds that are broken down and re-formed in protein turnover. Amino acid residues have molecular masses averaging about 100 g mol^{-1}, so 240 grams of protein contains about 2.4 moles of peptide bonds. Other experiments have shown that 9.6 moles of ATP (four molecules per bond) are needed to form 2.4 moles of peptide bonds. About 80 000 joules are needed to form a mole of ATP so the dog's daily energy requirement for protein turnover is about $9.6 \times 80\,000 = 770\,000$ J. By converting this rate per day to a rate per second, we find that it is 9 watts.

These very rough calculations suggest that of a 20 kilogram dog's 40 watt minimal metabolism, about 9 watts is used for pumping sodium ions, another 9 watts in protein turnover, and 5 watts to drive the heart. We seem to have explained about half the metabolic rate.

We have already seen that mammals use energy about five times as fast as similar-sized reptiles, even when their body temperatures are equal. A comparison of the metabolic rates of tissue slices from rats and lizards of equal mass, both at 37°C, gave a similar result. Rat liver and kidney use oxygen about four times as fast, per unit mass, as lizard liver and kidney; and rat brain uses oxygen twice as fast as lizard brain. Experiments with ouabain showed that sodium pumping in these tissues was using energy three to six times as fast in rats as in lizards. Further experiments showed that this was necessary to keep ion concentrations constant, because ions leak several times faster through rat cell membranes than through lizard cell membranes. Leaky membranes may seem wasteful of energy, but it seems possible that the benefits of endothermy could not be obtained without them.

Now we move on from mammals and lizards to a very different animal. The mussel *Mytilus edulis* is a common intertidal animal on both sides of the Atlantic. It feeds by pumping water through its mantle cavity and filtering out the diatoms and other unicellular algae. Its metabolic rate has been determined both by measuring oxygen consumption and by measuring heat production. It depends strongly on how much food is available. If there are plenty of suitable algae suspended in the water, mussels feed rapidly and have high metabolic rates. If no food is available, metabolism is much slower. Rates of protein turnover have been measured by feeding ^{15}N-labelled algae to mussels and observing the rate of appearance of radioactive ammonia in the water (young mussels, about 10 mm long, turned over 3.7 per cent of their protein daily). The energy cost of the turnover was calculated in the same way as for mammals and found to account for 20 per cent of metabolic rate when just enough food was given to prevent weight loss. This is very similar to what we concluded for dogs.

2.7 **Metabolism without oxygen**

Mussels are left high and dry at low tide. If they kept the valves of their shells open then, atmospheric oxygen could diffuse in but water vapour would diffuse out and the mussels would be in danger of drying up. Instead they clam up, closing their valves tightly so that water cannot escape, nor can oxygen get in. They may open their valves occasionally and admit a bubble of air, but their oxygen supply is very restricted until the tide rises and submerges them again, several hours later.

Animals can obtain energy from food both by aerobic metabolism (oxidizing the food in the usual way) or by anaerobic processes which make energy available without any need for oxygen. Measurements of oxygen uptake tell us only the rate of aerobic metabolism, but measurements of heat production can tell us the total rate of aerobic plus anaerobic metabolism. Simultaneous measurements of oxygen consumption and heat production of mussels have been used to find out how much use mussels make of anaerobic metabolism.

We have already seen that in aerobic metabolism, 20 joules of heat are produced for every cubic centimetre of oxygen used. In aerated water, the oxygen and heat production of mussels are in that ratio, showing that their metabolism is entirely aerobic. In air, heat production falls to about 15 per cent of the rate in water and oxygen consumption to about 5 per cent of the rate in water. Thus, mussels in air cut down their metabolic rate to 15 per cent of the rate in water and rely mainly on anaerobic metabolism. Mussels can also survive for several days in nitrogen gas, in which their metabolism has to be entirely anaerobic. When they are returned to aerated water after a period without oxygen, they use oxygen unusually fast for the first hour, but the extra metabolism then is far less than the energy saved by reducing the metabolic rate while out of water.

Mussels are short of oxygen only at low tide, but gut parasites such as tapeworms live permanently in an environment devoid of oxygen. They have to depend entirely on anaerobic metabolism. When kept in a laboratory in suitable solutions, they release lactate into the solution. It is produced by a sequence of reactions which are summarized by the equation

$$\underset{\text{glucose}}{C_6H_{12}O_6} = \underset{\text{lactic acid}}{2CH_3{\cdot}CHOH{\cdot}COOH} \tag{2.8}$$

If the solution contains dissolved carbon dioxide (which is plentiful in the natural host's gut) tapeworms also release succinate. This is apparently produced by the reaction

$$\underset{\text{glucose}}{C_6H_{12}O_6} + CO_2 = \underset{\text{succinic acid}}{(CH_2{\cdot}COOH)_2} + \underset{\text{pyruvic acid}}{CH_3{\cdot}CO{\cdot}COOH} + H_2O \tag{2.9}$$

(It does not seem to be clear what happens to the pyruvic acid.) These reactions yield far less energy than could be obtained by oxidizing the glucose. The enthalpy of combustion of glucose is -2803 kJ mol^{-1} and that of lactic acid is -1367 kJ mol^{-1} (Table 2.1). Thus the combined enthalpy of combustion of the two moles of lactic acid produced in eqn 2.8 is little less than that of the glucose, and the energy yield must be correspondingly small. Similarly, the enthalpies of combustion of succinic and pyruvic acids total little less than that of glucose, and the reaction shown in eqn 2.9 cannot yield much energy. Metabolism of a mole of glucose yields 38 moles of ATP if it is oxidized as in eqn 2.3; two moles of ATP if lactic acid is produced as in eqn 2.8; and three moles of ATP if succinate is produced as in eqn 2.9.

A tapeworm simply discards the fatty acids that its metabolism produces; because it cannot oxidize them, it can gain no energy from them. That is no great loss, in an environment like a host's gut in which foodstuffs are generally plentiful. Indeed, it would generally be of little advantage to a tapeworm to have an efficient metabolic process.

Mussels kept out of water can survive for at least eight days at 13°C. During this time they accumulate succinate in their bodies, together with smaller quantities of propionate (propanoate). Both are products of anaerobic metabolism, the succinate presumably produced by the same process as in tapeworms. While they are out of water, they have no opportunity to get rid of these products. When they are immersed again, they could excrete the succinate and propionate into the water, but a more economical alternative is to get more energy by oxidizing them. Energy released by oxidizing a fraction of the products is used to convert the rest back to the original foodstuffs. This happens during the burst of unusually fast oxygen consumption, after a period out of water.

This chapter has explained how metabolic rates are measured. It has illustrated the general rule that related animals have minimal metabolic rates proportional to (body mass)$^{0.75}$ and has presented the best explanation that has been devised up to now. It has also shown that animals of equal mass from different groups, for example a mammal and a fish, may have very different metabolic rates. It has shown that the energy costs of ion pumping and protein turnover account for a large fraction of the minimal metabolic rate. And it has discussed the anaerobic metabolism of gut parasites, and of molluscs left high and dry at low tide.

3 Feeding

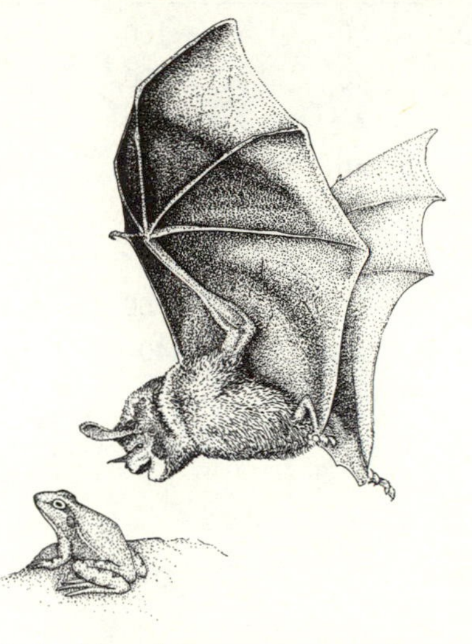

3.1 Herbivores and carnivores

Plants capture energy from solar radiation, herbivores get their energy supplies by eating plants, and carnivores get theirs by eating herbivores or, in some cases, other carnivores. We will start this chapter by looking at this energy flow in two ecosystems, a grassland and an ocean. Then we will consider how the guts of carnivores and herbivores are adapted to break down different kinds of food and make their energy available. Then we will examine the energy costs of feeding. Finally, we will discuss how animals should choose what they eat, to get as much energy as possible.

Our first example of an ecosystem is the Serengeti, 25 000 km^2 of grassland that is mainly in Tanzania but extends into the Mara region of Kenya. The most plentiful plants are grasses, but there are also stands of *Acacia* and *Commiphora* trees. The grasses are eaten by ungulate mammals, principally buffalo, wildebeest, zebra, and Thomson's gazelle. Also, some grass is eaten by rodents and grasshoppers, and a great deal is destroyed in frequent grass fires. Detritus (remains of dead plants) is eaten by termites, or used by them for building their mounds. The ungulates are eaten by large carnivores, principally lions and hyenas, and the rodents and grasshoppers are eaten by jackals. Hyenas kill much of their own food, but also scavenge on carcasses of animals that have been killed by lions or (more important) have died for other reasons. Vultures soar over the Serengeti, finding and scavenging carcasses in enormous quantities. Flesh-fly maggots and beetles also consume carcasses, and what remains is decomposed by bacteria. Dung beetles depend principally on the dung of the ungulates, and parasites are nourished by food eaten by their hosts.

Table 3.1 lists some of the more plentiful animals. Numbers of the larger

Table 3.1 Important animals of the Serengeti ecosystem

	Number/km²	Mass (kg)*	Biomass (kg/km²)
Wildebeest, *Connochaetes taurinus*	30	100	3000
Zebra, *Equus burchelli*	10	200	2000
Buffalo, *Syncerus caffer*	4	400	1600
Thomson's gazelle, *Gazella thomsoni*	40	15	600
Other ungulates			800
Total ungulates			8000
Lion, *Panthera leo*	0.1	150	15
Hyena, *Crocuta crocuta*	0.12	40	5
Other large carnivores			2
Total large carnivores			22
Jackals, *Canis* species		6	
Rodents			100
Vultures, *Gyps* species, etc.		2–8	10?
Grasshoppers			200
Termites			2000

*Masses are intended to be averages for both sexes and juveniles.
The sources of data are the same as for Fig. 3.1(a).

animals are from aerial surveys in the 1970s. Rodent populations were estimated from the results of trapping and grasshoppers were counted by a mark–release–recapture method; a sample were caught, marked, and released, then the size of the population was estimated from the proportion of marked individuals found in subsequent catches. Notice that the mass of carnivores in the table is only a tiny fraction of the mass of herbivores.

Figure 3.1(a) shows how matter flows through the ecosystem as one organism eats another. The rate at which the grass grows was measured by fencing an area off so that herbivores could not graze there, then cutting and weighing the grass at monthly (or longer) intervals. The rate of destruction by fire was estimated from the areas burnt over a few years. The masses of grass eaten by the various herbivores were estimated from measurements either of rates of food consumption or of metabolic rates of captive animals. The masses of prey eaten by carnivores and the total mass of herbivores dying each year were estimated from field observations. All the numbers in the diagram are rough estimates, and some are based on flimsy evidence. Many minor pathways are omitted, to keep it simple. Nevertheless, the picture it presents is probably broadly correct. Notice that the mass of food eaten by carnivores and scavengers is only a tiny fraction of the food intake of herbivores. The herbivores use or lose a lot of the mass and energy of their food in their metabolism and in their faeces. Only the matter and energy that the

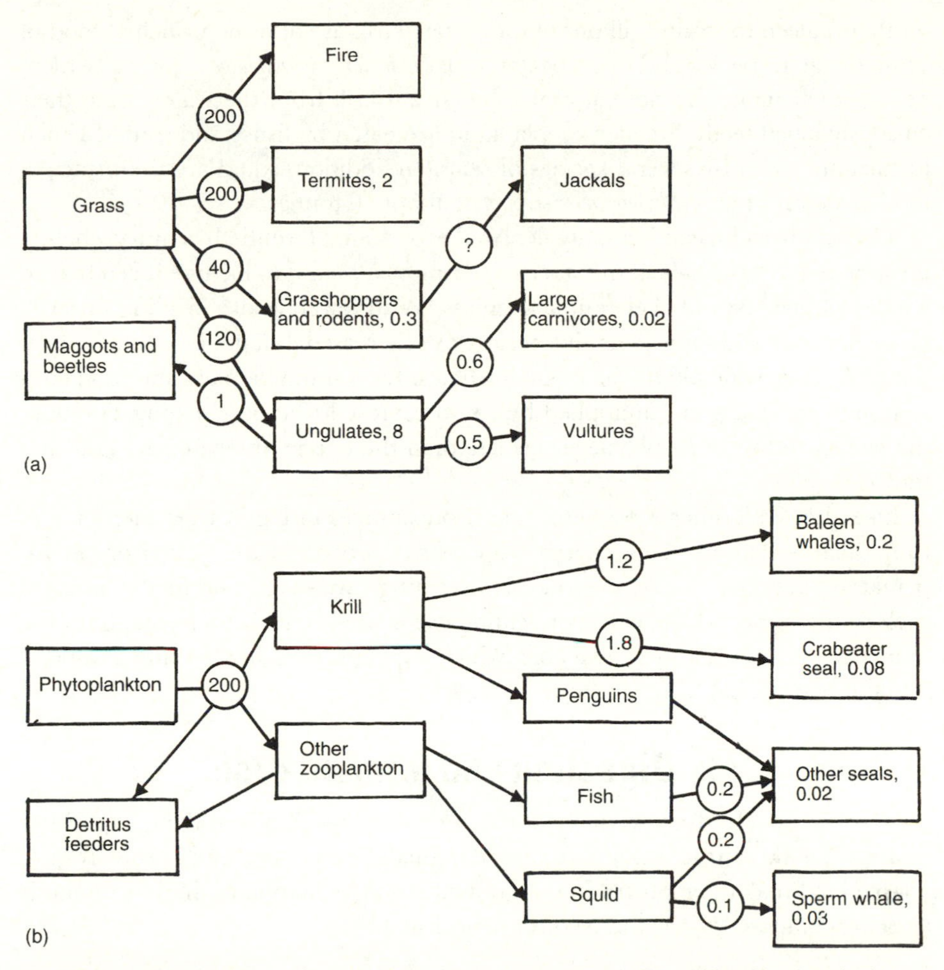

Fig. 3.1 Diagrams of the flow of matter through two ecosystems in East Africa. Numbers in the boxes are population biomass, expressed in tonnes per square kilometre. Numbers on arrows (in circles) show the rates at which matter is consumed, in tonnes per square kilometre per year. (a) The Serengeti, based on data in A. R. E. Sinclair (1975) *Journal of Animal Ecology* **44**, 497–520 and A. R. E. Sinclair & M. Norton-Griffiths (1979) *Serengeti: Dynamics of an Ecosystem, University of Chicago Press, Chicago.* (b) The Southern Ocean, based on data in papers from a symposium published in *Philosophical Transactions of the Royal Society* B **279,** 1–288 (1977).

herbivores are able to incorporate in their bodies as they grow and reproduce becomes available to the carnivores and scavengers.

Figure 3.1(b) is a similar diagram for the Southern Ocean, an area of about 36 million km^2 surrounding Antarctica. The phytoplankton are eaten by zooplankton, which, as well as the copepods and other small animals that are plentiful in other seas, include huge quantities of krill, shrimp-like crustaceans that are typically around 5 cm long. The krill are the principal food of the baleen whales, among which fin whales (*Balaenoptera physalus*, about 50 tonnes) and minke whales (*B. acutorostrata*, 7 tonnes) are the most numerous. These whales use the long fringes on their baleen to strain krill out of the water. Krill are also the principal food of some penguin species and of crabeater seals (*Lobodon carcinophagus*, 200 kg) which, despite their name, do not eat crabs, but strain krill from the water using their curiously lobed teeth. Smaller zooplankton are eaten by fishes and squids, which in turn are eaten by several species of seals. In addition, squid are the principal food of sperm whales (*Physeter macrocephalus*, about 30 tonnes).

The Southern Ocean is a sadly depleted ecosystem. Twentieth century whaling is estimated to have halved the stock of sperm whales and to have reduced baleen whales to one sixth of their initial biomass. With fewer whales feeding on krill, crabeater seals and some penguin species have increased, but not by enough to eat the krill made available by the missing whales. Even if the biomass and food consumption of whales are multiplied by six to represent the pre-whaling situation, the energy intake of the whales is far less than the energy intake of the krill that they eat.

Like almost all other ecosystems, the two examples in Fig. 3.1 get their energy from solar radiation. This energy passes first through plants (grass or phytoplankton), then herbivores, then herbivore-eating carnivores, and finally (in cases such as the sperm whale) carnivore-eating carnivores. The further you go up the sequence, the less energy is available, which explains (to quote the title of a book in the Bibliography) 'why big fierce animals are rare'.

3.2 **Digestion and fermentation**

In this and the next two sections we will look at the composition of animal and plant foods, and at how the guts of carnivores and herbivores are adapted to break them down. In this context it will be helpful to think of all foods as being composed of the three constituents listed below.

1. *Digestible materials* that are easily digested by the animal's enzymes. They may be carbohydrates, fats, or proteins.
2. *Fermentable-only materials* that cannot be digested by the animal's own enzymes but can be broken down to digestible products by symbiotic microorganisms in the animal's gut. These include materials, such as cellulose, which are major constituents of plant cell walls.
3. *Indigestible materials* that may be organic compounds such as lignin (in cell walls of woody stems) that neither the animal's enzymes nor those of its symbionts can break down, or inorganic compounds such as the calcium carbonate of coral skeletons and mollusc shells. These materials have no value as energy sources to animals.

Many of the foods eaten by herbivores contain a large proportion of fermentable-only cell-wall materials. Notable examples are the mature grass leaves eaten by cattle and other large ungulates and the dead or rotten wood eaten by termites. Ungulates have symbiotic bacteria and ciliate protozoans in their guts, which ferment cellulose, and termites have bacteria or flagellate protozoans. The very low oxygen concentrations in the gut ensure that the symbionts cannot oxidize the cellulose and gain all its enthalpy of combustion. Instead they break it down anaerobically into fatty acids, gaining a smaller quantity of energy and leaving the fatty acids for the host to absorb and metabolize. Thus the host and the symbionts share the available energy.

In cattle, carbohydrates are fermented by reactions such as

$$\underset{\text{glucose}}{C_6H_{12}O_6} \Rightarrow \underset{\text{acetic acid}}{2CH_3COOH} + \underset{\text{methane}}{CH_4} + \underset{\text{carbon dioxide}}{CO_2} \quad (3.1)$$

The enthalpy of combustion of glucose is –2803 kJ mol^{-1}, that of *two* moles of acetic acid is –1749 kJ, and that of methane is –890 kJ mol^{-1}. Thus the energy per mole of glucose that this reaction makes available to the microbes in the gut is 2803 – 1749 – 890 = 164 kJ, only 6 per cent of the enthalpy of combustion of the glucose. The methane is lost (cows burp) and its energy benefits neither the microbes nor the cow, but the cow gets the benefit of the acetic acid, with 62 per cent of the enthalpy of combustion of the glucose. Some other fermentation reactions produce less methane and yield more energy as fatty acids. In Section 3.4 we will assume a typical yield of fatty acids, containing 75 per cent of the enthalpy of combustion of the fermented food.

3.3 **Design of guts**

Digestion and fermentation are chemical processes, broadly similar to industrial processes in chemical engineering. In this section, principles from chemical engineering will be used to throw light on the design of guts. Animals' guts consist partly of wide chambers like mammalian stomachs, and partly of narrow tubes like small intestines. Similarly in chemical engineering, some processes are carried out in wide reaction vessels and some in narrow tubular reactors.

We have to consider these two types of reactor. In each case we will have a reagent which has concentration C_{in} as it enters the reactor, remains in the reactor where a chemical reaction occurs for time t, and leaves with concentration C_{out}. The rate constant for the reaction is r (which means that in a part of the reactor where the concentration of reagent is C, the reaction proceeds at a rate rC). A continuous-flow stirred tank reactor (CSTR) is a wide chamber whose contents are kept well stirred. A reagent entering the reactor is immediately diluted to the concentration at which it will leave. The concentration of the reagent is C_{out} throughout the well-mixed contents of the reactor, and the reaction proceeds at a rate rC_{out}. The change in concentration during the time t that the average reagent molecule remains in the reactor is

$$C_{in} - C_{out} = rC_{out}t$$

whence

$$C_{out} = C_{in}/(1 + rt) \quad (3.2)$$

In contrast, a plug-flow reactor (PFR) is a long narrow tube through which the reagents flow with very little mixing. At the entry end the concentration is still C_{in} and the reaction proceeds at a rate rC_{in} but as the reagent travels along the tube its concentration falls gradually to C_{out} and the reaction rate falls to rC_{out}. The fall is exponential and the final concentration as the reagent leaves the reactor is

$$C_{out} = C_{in}e^{-rt} \tag{3.3}$$

where e is the base of natural logarithms, 2.72. This tells us that reactions go faster in PFRs than in CSTRs. For example, when $rt = 1$, C_{out}/C_{in} is 0.5 for a CSTR (from eqn 3.2) but 0.37 for a PFR (eqn 3.3); 50 per cent of the reagent is broken down in a CSTR and 63 per cent in a PFR.

That suggests that PFRs should always be preferable, but they have a very serious limitation. If a PFR were used for a reaction performed by microorganisms (such as fermentation), new microorganisms would have to be added continuously. This is because a microorganism in a PFR is inevitably carried along it and out at the end. If no new microorganisms are supplied, the original ones will soon be washed out of the system. In a CSTR, mixing ensures that however many microorganisms are washed out of the reactor, some remain in it. If these remaining microorganisms reproduce fast enough to replace the ones that are lost, the population will be maintained. There will be no need to keep supplying new microorganisms.

The disadvantage of CSTRs, that they work more slowly than PFRs, can partly be overcome by arranging several CSTRs in series. Suppose that instead of one CSTR in which the reagents spend time t we have two CSTRs, each of half the volume, and that the reagents spend time $0.5t$ in each. If the concentrations leaving the two CSTRs are $C_{out,1}$ and $C_{out,2}$, it follows from eqn 3.2 that

$$\begin{aligned} C_{out,1} &= C_{in}/(1 + 0.5rt) \\ C_{out,2} &= C_{out,1}/(1 + 0.5rt) \\ &= C_{in}/(1 + 0.5rt)^2 \end{aligned} \tag{3.4}$$

(Note that the input concentration for the second CSTR is the output concentration from the first.) Equation 3.4 tells us that when $rt = 1$, 56 per cent of the reagent is broken down. This is a better result than for a single CSTR. A chain of more than two CSTRs would do even better, though still not quite as well as a PFR.

The position we have reached is as follows:

- digestible food can be digested or fermented, and digestion is preferable because the animal gets a larger energy yield;
- fermentable-only food can only be fermented;
- a PFR is better than a CSTR for digestion;
- for fermentation, only a CSTR will do;
- a series of CSTRs is better than a single one of the same total volume.

These considerations suggest that animals with no fermentable-only foods in their diets (for example, most carnivores) should have guts consisting solely of narrow digestive tubes which function as PFRs, and that those whose diets include a lot of fermentable-only food (most herbivores) should have guts including both PFRs in which digestible foods are digested and CSTRs in which fermentable foods are fermented. Fermentation and digestion cannot generally occur in the same part of the gut, because a fermentation chamber must not contain enzymes that would destroy the symbiotic microorganisms.

However, many carnivores eat their food in large chunks which need a capacious stomach to accommodate them for the initial stages of digestion, until they are broken down to smaller fragments. Most predatory fishes and all snakes swallow their prey whole, and cats and dogs swallow chunks of meat. Digestion starts in a wide stomach which probably functions as a CSTR, but then continues in a slender intestine which is clearly a PFR (Fig. 3.2(a)).

Herbivores generally have a PFR (an intestine) in which digestion occurs, but many also have one or more fermentation chambers which are (necessarily) CSTRs (Fig. 3.2(b),(c)). In cattle and other ruminant mammals, the principal fermentation chamber is the rumen, a modified part of the stomach, but there is also a smaller fermentation chamber in the hindgut. Sloths, kangaroos, and some leaf-eating monkeys also have modified stomachs that serve as fermentation chambers. In horses, however, fermentation occurs only in the posterior part of the gut, in a large intestine which is essentially tubular but is apparently wide enough to function as a CSTR, or as a series of connected CSTRs. It seems unlikely that food at the proximal end of the large intestine gets mixed well with food all the way along to the distal end, so the large intestine probably functions more like a series of CSTRs than a single one—we have seen how that can give an

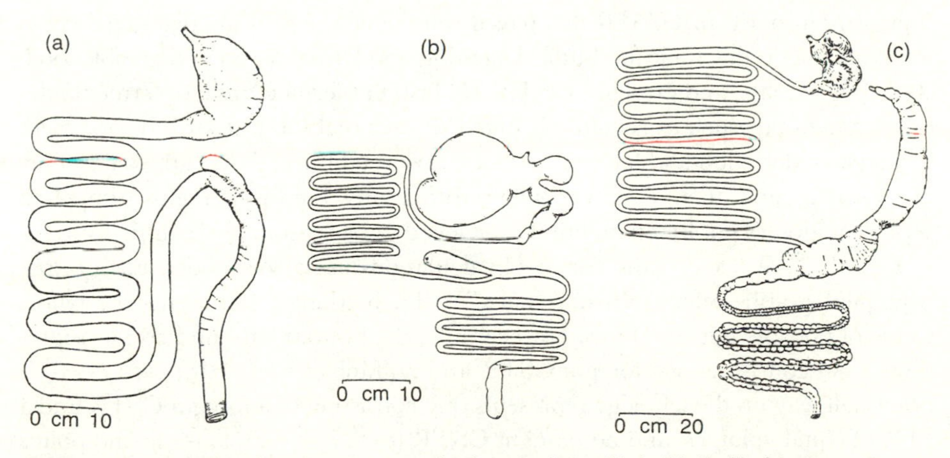

Fig. 3.2 Some examples of mammalian guts: (a) dog, *Canis familiaris*; (b) dik-dik, *Madoqua guentheri*, fermentation occurs principally in the rumen; (c) beaver, *Castor canadiensis*, fermentation in the caecum. (From C. E. Stevens (1988) *Comparative Physiology of the Vertebrate Digestive System*, Cambridge University Press.)

advantage. Rabbits and rodents also have fermentation chambers in the hindgut, but the important chamber in them is the caecum. Outside the mammals, fermentation occurs in the hindgut in herbivorous birds such as the ostrich and in reptiles such as *Iguana*, and even in a seaweed-eating fish (*Kyphosus*). Termites have their fermenting microorganisms in the hindgut. Thus fermentation chambers may occur at either end of the gut.

It is not immediately obvious that either arrangement is better than the other. If the food is sent through a PFR first, the wastage of energy that occurs when digestible foods are fermented will largely be avoided. However, there is a disadvantage in having a fermentation chamber at the anus end of the gut. The microorganisms that pass out of the chamber will be lost in the faeces; whereas if a digestive tube follows the fermentation chamber, microorganisms leaving the chamber can be digested and the animal can benefit from them. In an attempt to resolve the problem, I devised a computer model.

3.4 **A model of gut function**

The model is shown in Fig. 3.3(a). Its total volume is V, consisting of a CSTR of volume $v_1 V$, followed by a PFR of volume $v_2 V$, and a second CSTR of volume $v_3 V$. Each of the fractional volumes v_1, v_2, and v_3 can be given any value between zero and one, so the model can be made to represent a gut with a foregut fermentation chamber, or a hindgut fermentation chamber, or both, or neither. Food enters at the mouth end at a chosen rate. It is fermented in each of the CSTRs and digested in the PFR, as it passes through them. It will be assumed that fermentation produces fatty acids containing 75 per cent of the enthalpy of combustion of the fermented food and microorganisms containing 10 per cent; the rest is used in the microorganisms' metabolism, or lost as methane. Microorganisms from the first CSTR that pass through into the PFR are digested there at the same rate as digestible foodstuffs. Digestion and fermentation of digestible foodstuffs (sugars, fats, and proteins) is relatively fast, but fermentation of fermentable-only compounds such as cellulose is much slower; realistic rates will be assumed. Products of digestion and fermentation are absorbed through the gut wall as they are formed, and their energy content totted up. This is a highly simplified representation of gut function, but it is believed to be reasonably realistic.

Figure 3.3 also shows the results of calculations using the model. Each of the triangular graphs refers to a different diet. Each triangle represents the whole range of possible gut structures. For example, the bottom left hand corner represents a gut consisting of a foregut CSTR and nothing else ($v_1 = 1$, $v_2 = v_3 = 0$); a point half way up the left edge represents one consisting of a foregut CSTR and a PFR of equal volumes, and no hindgut CSTR ($v_1 = v_2 = 0.5$, $v_3 = 0$); and points inside the triangle represent guts having all three components. The contour lines show how much energy the animal would gain from the food, for all possible gut structures.

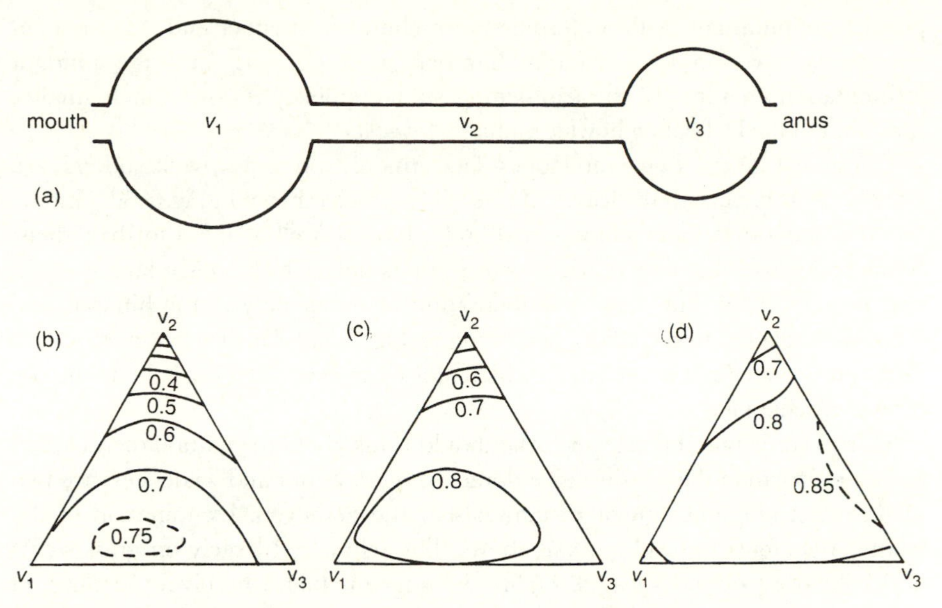

Fig. 3.3 (a) The model gut represented in computer simulations. (b), (c), (d) Graphs showing calculated rates of energy gain by a range of possible guts, for three different diets. Each triangle represents all possible combinations of the fractions v_1, v_2, v_3 that describe the range of possible gut designs, with contours showing the fractions of the diet's energy that the animal gains. The three triangles refer to diets of the following compositions (expressed as fractions of total energy content):

	Digestible	Fermentable-only	Indigestible
(b) Poor diet	0.05	0.50	0.45
(c) Moderate diet	0.30	0.50	0.20
(d) Rich diet	0.50	0.30	0.20

From R. McN. Alexander (1993) *Journal of Zoology* **231**, 391–401.

Figure 3.3(b) refers to a poor diet such as mature grass or mature leaves of trees, containing a very small proportion of digestible material. The contours show that the gut that extracts most energy from this food has $v_1 = 0.6$, $v_2 = 0.1$, $v_3 = 0.3$; it has a large fermentation chamber in the foregut and a smaller one in the hindgut, with a small digestive intestine between them. The guts of cattle, antelopes, and other ruminants are like this. Figure 3.3(c) refers to a richer diet with a higher proportion of digestible material, designed to resemble young leaves. All guts extract more energy from this than from the poorer food, but the optimum gut is again ruminant-like. Finally, Fig. 3.3(d) refers to a gut that is richer still, though not as rich in digestible material as some grains and fruits. For this diet, the optimum gut has $v_1 = 0$, $v_2 = 0.3$, $v_3 = 0.7$; it has no fermentation chamber in the foregut, but a large one in the hindgut. Horses, rabbits, and rodents have guts like this. Calculations for diets consisting almost entirely of digestible material (for example, soft fruits containing high proportions of sugars, or the flesh of animals) predict optimum guts with no fermentation chamber at either end. Mammalian carnivores have no fermentation chambers in their guts, and the hindgut fermentation chamber of the fruit-eating spider monkey *Ateles* is much smaller than that of the leaf-eating howler monkey *Alouatta*.

Cattle eat grass. Some antelopes feed mainly on grass while others are principally browsers on the leaves of trees. The results shown in Fig. 3.3(b) and (c) seem to show that these animals have guts that are well adapted to their diets. Some rodents feed mainly on rich foods such as nuts, which consist largely of fat and protein. Their guts have a fermentation chamber only in the hindgut and seem well adapted to such diets, according to Fig. 3.3(d). However, some hindgut fermenters feed mainly on grass, which does not seem at first sight to fit the model's predictions.

Horses come into this category. We should think about the natural diets of wild horses rather than the diets we give domestic ones. Zebra and wildebeest are two of the most numerous grazing mammals in the Serengeti. Examination of the stomach contents of dead animals shows that zebra feed largely on grass stems and wildebeest on grass leaves. Figure 3.3 suggests that a herbivore feeding on poor food such as grass will do best if it has a large foregut fermentation chamber, which wildebeest (like other ruminants) have. Grass stems seem even poorer food, suggesting that zebra also would do best with a rumen, but they (like other horses) have only a hindgut fermentation chamber—zebra seem adapted for richer food than wildebeest. It seems possible that the impression that zebra depend on poor food may be false. Grass stems are indeed poor food, but the grass seeds that grow at the tops of the stems are good, starch-rich food. The stems are much more noticeable than the seeds in the gut contents, but the seeds may make the diet much richer than at first appears.

However, we should avoid the assumption that all animals are optimally adapted to their diets. There may be no consistent differences of diet between wild horses and antelopes; it may just be a historical accident that horses have evolved from hindgut-fermenting ancestors, and antelopes from foregut fermenters. Both kinds of gut work pretty well for all kinds of plant diet, and it may be difficult for a group that has one gut arrangement to evolve the other.

Rabbits are also grazers with only a hindgut fermentation chamber, but they avoid a disadvantage of hindgut fermentation by the unpleasant habit of coprophagy; that is, by eating their own faeces. Remember that hindgut fermentation has the advantage that digestible materials do not get fermented (which would waste some of their energy); and it has the disadvantage that microorganisms passing out of the fermentation chamber are lost in the faeces, giving the animal no opportunity to digest them and so capture their energy content. Rabbits and many rodents eat some of their faeces, and so digest the microorganisms that would otherwise be lost. They feed at night, and it is during the day (when they are resting) that they re-ingest faeces. The faeces are almost inevitably a less good food than grass (most of the easily digested or fermented material in the diet has been removed) but if faeces are eaten at a time of day when the resting animal would otherwise be eating nothing, they may be a useful source of energy.

Hares (which eat grass) and possums (*Pseudocheirus*, which eat eucalyptus leaves) habitually practise coprophagy. Experiments in which they have been prevented from eating their faeces have shown that the habit may increase the energy they extract from their food by 10–20 per cent. No foregut fermenters are known to be coprophagous, and calculations using the same model as Fig. 3.3 indicate that they would gain very little from the habit. Most of their microorganisms get digested in the intestine, and few reach the faeces.

Cattle spend a great deal of their time chewing the cud, about eight hours per day. While they are chewing they cannot eat, and if they did not chew they would have time to eat a great deal more. But chewing breaks the food down into smaller fragments which can be expected to be fermented faster and more completely, because a bigger area is exposed to attack by gut microorganisms. Experiments have confirmed that finely ground food is indeed fermented faster. It will also be digested faster.

A cow that did not chew would lose a lot of undigested grass in its faeces. One that chewed all day would have no time to feed. There must be an optimum fraction of the time that should be spent chewing. Calculations using the chemical reactor model of the gut (Fig. 3.3) confirm, as might be expected, that animals that eat poor food (containing a low proportion of digestible material) should spend longer chewing than those that eat rich food; and that animals feeding on food that is resistant to chewing should chew longer than those whose food breaks down easily. These theoretical predictions fit the observations that cattle and other grazing animals spend a lot of time chewing, but carnivores such as lions and domestic cats bolt big chunks of food. Birds have no teeth and cannot chew, but herbivorous birds such as ostriches and domestic fowl swallow stones and grit which are held in a muscular gizzard. Food passing through the gizzard is ground between the stones.

Most mammals do all the chewing they are going to do before the food leaves the mouth. Ruminants chewing the cud return partly fermented food to the mouth for further chewing. This is easily done for animals with a foregut fermentation chamber but would not be practicable for hindgut fermenters such as horses. It has the advantage that chewing need not be done immediately, but can be deferred to a convenient time.

3.5 Costs of feeding

We have been discussing the energy content of the products of digestion that are absorbed through the gut wall. We have seen how it depends on diet and gut structure, but so far we have ignored an important point—the energy absorbed through the gut wall is not all gain to the animal. Both feeding and digestion are processes with energy costs that must be paid out of the energy that they bring to the animal. We will start by examining these costs in mussels, which have been the subject of research that establishes the principles particularly well. Then we will look briefly at sheep and trout. In all three cases we will find that the net energy gain from feeding is substantially less than the enthalpy of combustion of the absorbed products of digestion.

If an animal that has been starving is given food, its metabolic rate rises. It remains elevated while digestion continues, for some time after the end of the meal. The extra heat liberated by metabolism, over and above the metabolism that would have occurred if the animal had remained starving throughout, is best called the **heat increment of feeding**. It is often called the specific dynamic effect, but I prefer to avoid that term because there is nothing peculiarly specific or dynamic about it.

Figure 3.4(a) shows what happens when mussels that have been starved for a long period are fed for a few weeks. Their metabolic rate (measured as heat production) begins to rise as soon as feeding starts and levels off after a few days. It remains high while feeding continues, but when feeding is discontinued the metabolic rate returns gradually to its original level. The shaded area in Fig. 3.4(a) represents the heat increment of feeding.

Energy is needed for feeding, for digesting and absorbing the food, and for the processes of growth. We will try to find out how much of the heat increment relates to each.

Mussels feed by driving water over their gills and filtering out any unicellular algae or other food particles that are suspended in it. The water current is driven by cilia which are active so long as there are particles suspended in the water. When there are no particles, the cilia beat only intermittently, driving much less water over the gills, but still enough to keep the animal supplied with oxygen.

There must be an energy cost of feeding, because energy is needed to keep the cilia beating continuously. An attempt was made to measure this cost by an experiment with a suspension of polystyrene spheres, of about the same diameter as the algae on which mussels usually feed. Mussels that had been resting in particle-free water were immersed in the suspension. Their cilia became fully and continuously active, driving water rapidly over the gills and filtering out the spheres (which were then discarded without being swallowed), but there was no detectable rise in metabolic rate. It seems that, for mussels, the energy cost of feeding is very small.

We return now to real feeding, with algae. Figure 3.4(a) shows the metabolic rate increasing sharply when feeding starts, during the few hours it took to fill the gut. It continues to increase, but more slowly, over the next few days, until eventually a steady state is reached. When feeding ends, the metabolic rate falls sharply during the few hours it takes to empty the gut again, then more slowly until it gets back to the starvation level. The sharp fall d is about equal to the sharp rise d that occurred while the gut was filling. These observations suggest that d is the energy cost of digesting and absorbing the food, and that the longer term

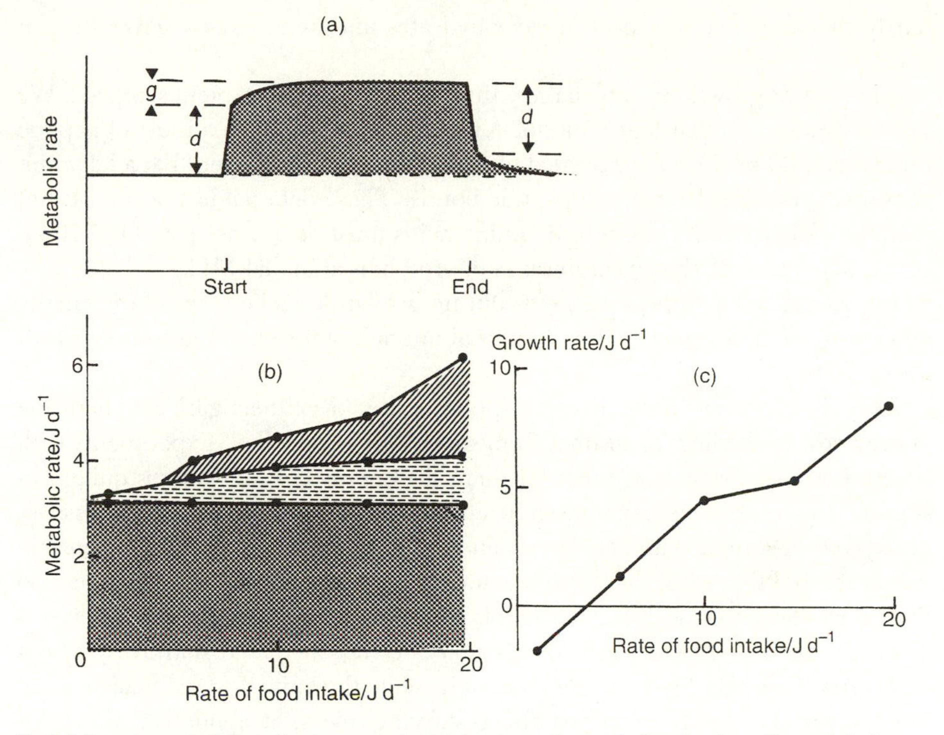

Fig. 3.4 Energy cost of digestion in mussels. (a) A schematic graph showing how the metabolic rate of a mussel changes during and after a meal. *d* is interpreted as the energy cost of digestion and *g* as the cost of growth. (b) A graph of steady-state metabolic rate against rate of food intake, showing the components associated with different functions. The lowest tinted band represents the maintenance requirement, the metabolic rate when not feeding; the middle band represents the additional metabolic rate *d* attributed to the costs of digestion and absorption; the top band shows the component *g* interpreted as the cost of growth; and the line at the top is the total metabolic rate. (c) A graph of growth rate against rate of food intake. In these graphs, food intake is expressed as metabolizable energy intake, that is as ($H_{food} - H_{faeces} - H_{urine}$) in eqn 2.2. (Data for (b) and (c) are from J. Widdows and A. J. S. Hawkins (1989) *Physiological Zoology* **62**, 764–84.)

change *g* represents the energy cost of incorporating surplus food in the body as growth.

Figure 3.4(b) shows results of experiments in which mussels were given algal suspensions of different concentrations and accordingly took in food at different rates. Metabolic rates were measured sufficiently long after a change of suspension for a steady state to have been reached. Starving animals metabolized at the minimum rate. Costs of digestion and absorption, and of growth, were measured as shown by *d* and *g* in Fig. 3.4(a). Rates of growth are shown in Fig. 3.4(c). At low feeding rates, growth is negative. At the highest rate of food intake, the cost of digestion and absorption is 5 per cent of the rate of food intake, and the cost of growth is 25 per cent of the rate of growth. The cost of digestion and absorption is probably partly the cost of the protein turnover involved in enzyme secretion, and partly that of active transport of carbohydrates and amino acids across the gut wall.

The cost of growth must be largely that of the necessary protein synthesis. We saw in Section 2.6 that four moles of ATP are needed to form a mole of peptide bonds; that 80 kilojoules are needed to make a mole of ATP; and that a kilogram of protein contains 10 moles of peptide bonds. These data tell us that the cost of making a kilogram of protein from amino acids must be at least $4 \times 80 \times 10$ kJ, or 3.2 MJ. The enthalpy of combustion of protein is about 24 MJ kg^{-1} (Table 2.1), so the cost of forming peptide bonds during growth is 13 per cent of the energy content of the new protein. This explains about half of the observed cost of growth for mussels.

Mussels need very little energy to pump water over their gills; for them the energy cost of feeding (as distinct from digestion) is very low. Experiments with sheep, however, have shown that for them the energy cost of eating is quite substantial. In one set of experiments, a sheep was kept in a calorimeter while its rates of oxygen consumption and heat production were measured. The oxygen measurements showed that while the sheep was standing in the calorimeter, having had no food since the previous day, its metabolic rate was about 100 watts. When it was given chopped dried grass to eat, its metabolic rate increased sharply, to about 140 watts. This very high rate persisted only while the animal was actually eating, but the metabolic rate remained above starving levels, at about 120 watts, for some hours after the meal.

The energy cost of digesting and absorbing food can slow well-fed animals down. In experiments with young rainbow trout (*Onchorhynchus*), fish given as much food as they would eat had resting metabolic rates 68 per cent higher than starving fish and could not swim as fast. The maximum speed that they could sustain for a substantial distance was 15 per cent lower than for the starving trout. It appeared that sustained speed was limited by the rate at which the gills and blood system could supply oxygen to the muscles; at their maximum speeds, both groups of fish used oxygen at the same rate.

3.6 Quality or quantity?

We have discussed the effectiveness of different gut structures, for dealing with different diets, but we have not so far considered how animals should choose, if a variety of foods is available. Often, the choice is between foods of different qualities. Should animals eat all the food they can get, or concentrate on the best stuff? Should a bird eat all the worms it finds, or only the big juicy ones? Should a crab eat all the molluscs it finds, or only the ones whose shells it can break open easily? Because this is a book about energy, we will tackle these questions from a strictly energetic point of view, asking how the animal can maximize its energy intake. We will ignore the fact that food choice must sometimes be dictated by other considerations. For example, it may sometimes be more important for a growing animal to get enough protein than to get the greatest possible quantity of energy.

There is a simple theory that tries to predict how animals should choose diets to maximize energy gain. Consider an area containing n_1 first class prey, each of which has energy content E_1; and n_2 second class prey with energy content E_2. If it never stopped to feed a predator could search the area in time T, but it uses additional time t_1 every time it stops to pick up and swallow a first class prey item, and time t_2 for every second class item. The first class prey may be better either because it has the higher energy content ($E_1 > E_2$) or because its handling time is shorter ($t_1 < t_2$) or both. More precisely, the criterion that makes it better is that it has the higher ratio of energy content to handling time ($E_1/t_1 > E_2/t_2$), so energy can be got from it faster.

Taking only second class prey would obviously not be a sensible option, but taking only first class prey might be. If the predator does this it will spend time ($T + n_1t_1$) searching the area and eating all the first class prey, from which it will get energy n_1E_1. Its rate of energy intake will be $n_1E_1/(T + n_1t_1)$. Similarly if it eats all the prey of both kinds its rate of energy intake will be $(n_1E_1 + n_2E_2)/(T + n_1t_1 + n_2t_2)$. It should eat both kinds if

$$(n_1E_1 + n_2E_2)/(T + n_1t_1 + n_2t_2) > n_1E_1/(T + n_1t_1)$$

If you do a bit of algebra you will find that this means it should eat both kinds of food if

$$n_1 < E_2T/(E_1t_2 - E_2t_1) \tag{3.5}$$

This means that if there are plenty of first class prey (n_1 is large) it should eat only them, but that if first class prey are few and far between it should eat both kinds of prey.

A slightly more complicated argument shows that the best strategy is an 'all or nothing' one—it is never best to eat just some of the second class prey.

Experiments have been done to find out whether birds and crabs feed as this theory suggests they should. In one set of experiments, great tits (*Parus major*) were offered mixtures of two kinds of food, which were carried through their cage on a moving belt. Both kinds were pieces of mealworm, relatively large pieces and small ones only half their size. The small ones had pieces of sticky plastic tape attached, which the birds had to peel off before they could be eaten. Thus the smaller pieces had lower energy content and required a longer handling time; the ratio E/t was lower for them than for the large pieces, making them clearly the less profitable prey.

When the food was supplied at low rates (for example, three pieces per minute) the birds fed indiscriminately, eating every piece that was offered. However, when the big pieces were offered at high rates the birds concentrated on them, almost (but not quite) ignoring the small ones. The change from indiscriminate to selective feeding occurred at the theoretically predicted rates of food supply, but the birds continued to eat a few small pieces even when the theory said they should eat only large ones.

This was a highly artificial experiment and interpretation of it was complicated by the possibility that the birds may have been able to predict the time of arrival of the next large piece, and take account of this in deciding whether to start eating a small one. However, redshank (a wading bird, *Tringa totanus*) have been shown to behave similarly when feeding in their natural habitat. They use their long bills to catch worms by probing in mud-flats. On mud containing few worms they eat every worm they find but when large worms are plentiful they concentrate on them, and eat very few small ones. The large worms took longer to eat than the small ones, but not enough to negate the advantage of their larger size—they had larger values of E/t.

Crabs (*Carcinus maenas*) feed on mussels where these are available. They use their pincers to break the mussels' shells to get at the flesh inside. Small mussels contain little flesh but can be broken and eaten quickly, in a few minutes. Large ones contain much more flesh but may take an hour or more to break and eat. An intermediate size is the most profitable, giving the highest value of E/t for a given size of crab. In experiments in which they were supplied with a range of sizes of mussels, crabs at first ate mainly the most profitable size, but went on to eat larger and smaller mussels when the most profitable size was finished.

For herbivores the important difference between two foods is often not their energy content, but the proportion of digestible to fermentable-only material. For example, an antelope may have the choice of eating grass indiscriminately, in which case it will be able to eat fast; or of seeking out the succulent young shoots, in which case it will take in a smaller quantity of more digestible food. The relative merits of these two strategies have been compared, using the model illustrated in Fig. 3.3. This analysis showed that in circumstances in which even indiscriminate feeding has to be slow (for example, when food is sparse) there is plenty of time for cellulose to be fermented and it is best to eat everything available. However, when it is possible to eat fast it will often be better to feed selectively, although this means eating less. This may explain two observations. Grazing antelopes are less selective when feeding where the food supply is sparse. And large grazing animals such as buffalo are generally less selective than small ones such as gazelles. (Remember that larger animals have lower metabolic rates per unit mass, and take food in at lower rates per unit body mass.)

3.7 **Patches of food**

Now we ask another question about feeding. How should animals move around, as they search for food, to get as much energy as possible? Many foods are found in patches. For a tit feeding on insects, every bush can be thought of as a separate patch. The longer the bird remains on one bush, the fewer insects will be left for it to eat and the harder they will be to find. The bird should stay on each bush for some time, but the time will come when its best policy is to move to another. The move may be short and easy, but it may cost quite a lot of time and energy if the next bush is a long way off. Time and energy will be wasted if the tit moves too often.

As another example, think of a ladybird beetle larva feeding on aphids. It does not swallow them whole but eats the flesh from inside the hard cuticle. Each aphid can be thought of as a patch. The longer the ladybird larva stays on it, the less flesh is left and the harder it is to get the remainder. The larva's rate of food intake will fall until eventually its best policy is to abandon the carcass and set off in search of another aphid. It may find one soon, but it may face a long search.

We assume that the aim of a patch-feeding animal is to gain food energy as fast as possible. The rate that concerns us is the average rate, taking account not only of the periods of feeding, but also of the time taken to travel from one patch to the next, or to find a new patch. An animal that feeds in a patch for time t gains energy $E(t)$. A graph of $E(t)$ against t will not be a straight line but will curve as shown in Fig. 3.5(a), because as feeding continues the remaining food is progressively harder to get. Suppose that the average time taken to find a new patch is T. Then if the animal stays in each patch for time t and then moves on, its average rate of energy intake will be $E(t)/(T + t)$. The animal would like to make this rate as large as possible.

Figure 3.5(b) shows a convenient way to find the time t that does this. A point has been marked on the horizontal axis at time $-T$. A line from this point which

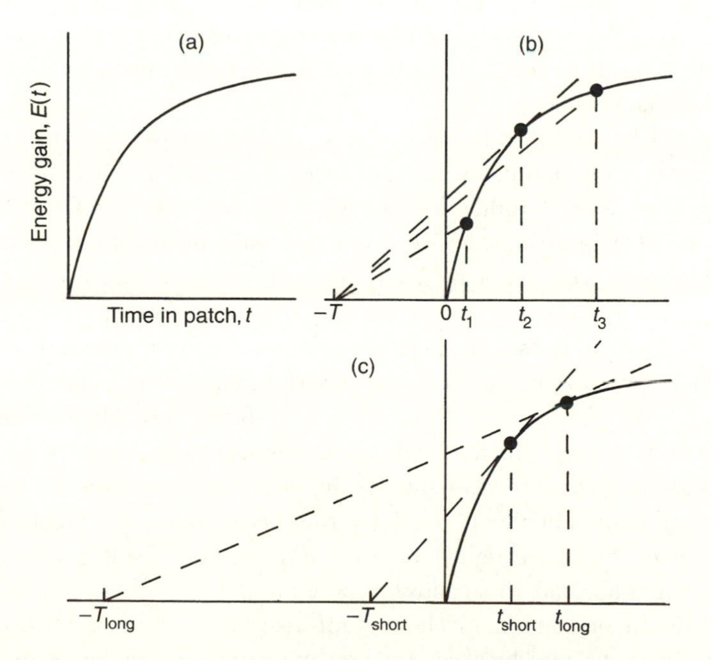

Fig. 3.5 Schematic graphs of energy gain, while feeding in a patch of food, against time. The text explains how graphs like these can be used to find the optimum time to move to another patch.

intersects the energy intake graph at time t has slope $E(t)/(T + t)$, which is the quantity we want to make as large as possible. The line of this kind that has the steepest slope is the tangent, which touches the curve at time t_2; notice, for example, that this line is steeper than the ones that cut the curve at times t_1 and t_3. This tells us that the animal's best strategy is to spend time t_2 in each patch before moving on.

Figure 3.5(c) shows that the longer it takes to travel to the next patch, the longer the move should be delayed. When mean travel time is T_{short} the optimum time in the patch is t_{short} (found by drawing the tangent), but when travel time is T_{long} optimum time in the patch is t_{long}.

In experiments designed to find out whether ladybird larvae behave as this theory suggests, a larva was put on a tray 0.5 metre square with 2, 4, 8, 16, or 32 aphids. It fed on one aphid for a while, then left it to search for another, and its movements were timed. Whenever it abandoned the remains of an aphid, the carcass was removed and a new aphid supplied, so that the number of uneaten aphids available remained constant. When there were 32 aphids on the tray, the larvae found new ones quickly (in about 15 minutes) and feeding time on each aphid was relatively short (25 minutes). With only two aphids on each tray it took longer (35 minutes) to find the next one and the larva stayed longer (50 minutes) on each. This is the kind of relationship between search time and feeding time that Fig. 3.5(c) leads us to predict. Graphs like Fig. 3.5(a) were obtained by giving larvae just one aphid each and measuring the rate of food intake by removing and weighing the remains of the aphid after various times. Optimum feeding times for various search times were predicted by drawing tangents; these agreed well with the feeding times that had been measured in the experiments in which several aphids had been supplied.

Another set of experiments was designed to discover whether tits hunting for insects behave in the optimal way. 'Trees' made of wooden dowel were set up in an aviary (they looked rather like leafless Christmas trees). To some of the branches, small cylindrical boxes were attached, each filled with sawdust in which a few mealworms were hidden. Ideally these boxes would have been placed at different distances apart so that substantially different times were needed to fly from one to the next. In fact, the aviary was far too small for that to be practicable so the time required to open a box was varied, instead of the time to travel to it. This was done by giving the boxes loose- or tight-fitting lids. The birds learned to remove the lids to get at the food, taking on average 5 seconds to remove a loose lid and 21 seconds to remove a tight one. In each experiment the lids were either all loose or all tight. The theory predicted that tits should spend longer feeding at each box when lids were tight, and they did; average feeding times were 74 seconds for tight lids and 46 seconds for loose ones.

To test the theory quantitatively, a graph like Fig. 3.5(a) was constructed; the information was obtained by observing how many mealworm pieces the tits found in different feeding times. It emerged that tits generally spent a little longer at each box than the simple theory predicted. A possible explanation is that the birds take account of the energy cost of removing the lid, as well as the time. The mean rate of energy gain is not exactly $E(t)/(T + t)$, as we have been assuming, but $[E(t) - M(T)]/(T + t)$, where $M(T)$ is the metabolic energy cost of removing a lid that takes time T to get off. When the rate of energy gain was calculated in this way, feeding times agreed well with the theoretical optima.

Now consider a new and very different example, bumble-bees collecting nectar from flowers. Plants such as the foxglove (Fig. 3.6) have their flowers arranged in spikes, and each flower spike can be thought of as a patch of food for bees. If all the flowers were identical, a bee's best strategy would be to go systematically from flower to flower, visiting every flower in one spike and then moving to the next. However, the flowers are not identical.

The arrangement that is found is the product of co-evolution of bees and flowers, each evolving to take advantage of the other. The bee confers a benefit on the plant by pollinating the flowers. If it transferred pollen from a flower to another on the same plant, the plant would suffer the disadvantages of inbreeding—in particular, deleterious recessive genes would be apt to become homozygous in its offspring, and so get expressed. The plant has evolved to avoid this by forming a flower spike on which buds open progressively from the bottom up. The freshly open flowers are male; their anthers are producing pollen but the stigma is not receptive. The older flowers, lower down the spike, are female; their stigmas are receptive and their anthers no longer bear pollen. The optimum course of events from the plant's point of view is for each bee to visit the female flowers first (bringing them pollen from other plants) and then move on to the male ones

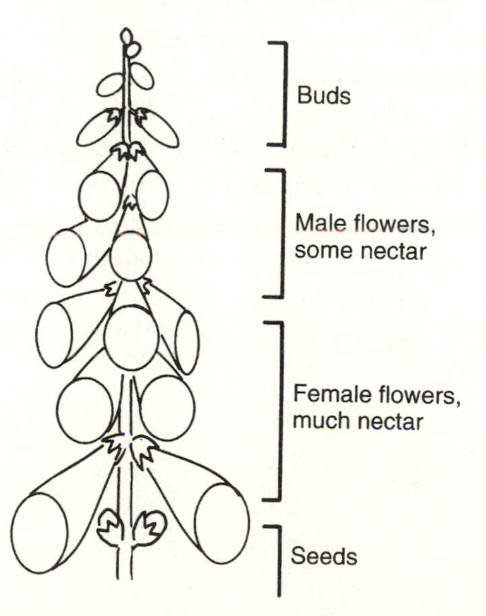

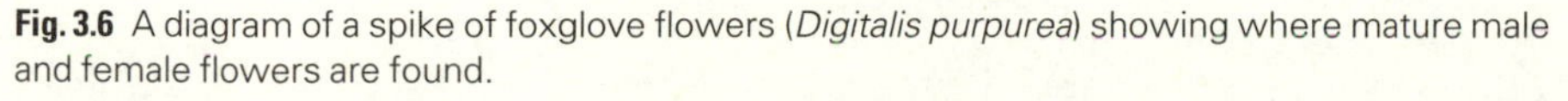
Fig. 3.6 A diagram of a spike of foxglove flowers (*Digitalis purpurea*) showing where mature male and female flowers are found.

(picking up pollen that will be carried to the next plant). The plant has evolved in such a way as to make this also the optimum strategy for the bee.

There is a gradient of nectar production up the spike. The oldest female flowers produce most nectar and the youngest male ones produce least. Visiting bees deplete the nectar which is replenished by the plant over the following 24 hours. A bee arriving at a plant does not know whether or when it has been visited by other bees, so cannot know whether any of the flowers contain enough nectar to be worth visiting. However, it does know from experience that the lowest flowers can be expected to contain most nectar. If any flower is worth visiting, the bottom one will be. The bee's best strategy is to start at the bottom of the spike and work systematically upwards. As it moves up it will get less nectar from successive flowers, so a graph of nectar intake against time will look like that shown in Fig. 3.5(a). Eventually, as the intake rate falls, the bee should leave that flower spike and move to another; the optimum time for the move can be predicted using a graph like Fig. 3.5(b). Bees were observed foraging on foxgloves in Washington State, where foxgloves have escaped from cultivation and become very common. The observations showed that the bees behaved more or less according to the theory. They started at the bottom flower of the spike more often than at any other. They worked systematically up the spike but commonly skipped some of the flowers, apparently because of a tendency to move vertically up the spike instead of spiralling round it. They left the spike at (approximately) the theoretically optimum time.

We started this chapter by looking at two ecosystems, seeing how solar energy captured by plants passes to the herbivores that eat them, then to herbivore-eating carnivores and, possibly, carnivore-eating carnivores. At every stage, energy is used in metabolism or lost in faeces, so that the carnivores gain only a tiny fraction of the energy captured by the plants. Then we considered the special problems of herbivores, which eat food containing large proportions of cell-wall materials that their enzymes cannot digest. We discussed the merits for different diets of different gut designs, with different arrangements of digestive tubes and fermentation chambers. Next we examined the energy costs of feeding, of digestion, and of growth. Finally, we asked how animals should forage for food, to gain energy as rapidly as possible; which kinds or sizes of food should they choose; and how long should they remain feeding in one patch of food before moving to the next?

4 Movement

This chapter is about the energy that animals use for travelling around; walking, flying, or swimming. For many of them, this is the principal energy cost additional to the basic costs discussed in Chapter 2. We will discuss the energy costs of moving on land, in the air, and in water, how these costs arise, and how animals adjust their patterns of movement to keep the costs as low as possible. What gaits should animals use, how should they plan their journeys, and when is a long migration worthwhile? We will also consider the mechanisms that make possible the bursts of very high power output that are needed when an animal sprints to catch prey or to escape a predator.

4.1 Walking and running

The energy costs of walking and running have generally been measured indirectly, by measuring animals' rates of oxygen consumption. A large part of the best research was done by the late Professor C. R. (Dick) Taylor of Harvard University, and I will describe the methods that he developed and used.

For most of their experiments, Taylor and his colleagues trained animals to run on treadmills (short conveyor belts designed for the purpose, Fig. 4.1). The animal walks or runs in the direction opposite to the movement of the belt, matching its speed to the belt's speed so as to remain stationary relative to the laboratory. This arrangement gives two important advantages—the animal's speed can be controlled by varying the speed of the belt, and the physiological equipment (which I will soon describe) does not have to be moved, to keep it alongside the animal. It is remarkably easy to train animals to run on the belt and it has been done successfully with animals ranging from mice to elands (*Taurotragus*, a large antelope). However, when the team worked with an elephant it seemed too expensive to

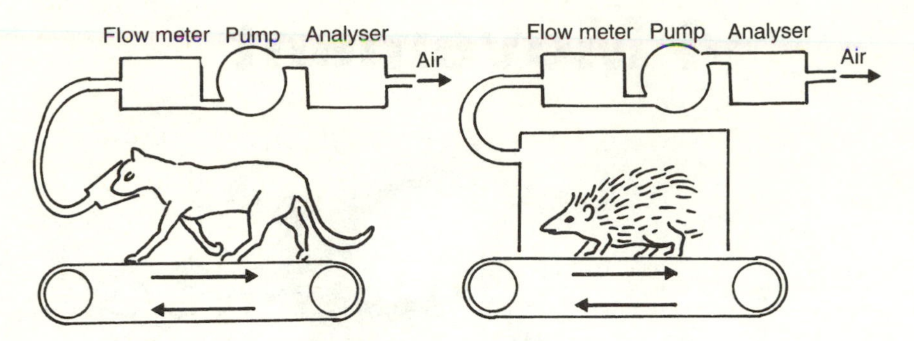

Fig. 4.1 Measuring the oxygen consumption of animals running on treadmills. In (a) the animal wears a mask from which air is sucked for analysis; in (b) the whole animal is enclosed by a hood from which air is drawn.

build a sufficiently strong treadmill and they put their equipment on a golf buggy which could be driven alongside the elephant as it walked around the zoo. It is very important that the animal should be thoroughly accustomed to the experiment because, if it is excited, its metabolic rate will be abnormally high.

The air that the animal breathes out must be collected and analysed to find out how much oxygen has been used. In experiments with reasonably large animals (the size of a cat, or larger) a loose-fitting mask is fixed over the nose and mouth (Fig. 4.1(a)). A pump draws air from the mask through a flow meter and an oxygen analyser. The loose fit of the mask allows air to leak in around its edges, keeping the animal supplied with fresh air. If this pump is working fast enough (and tests are done to make sure that it is) no air leaks out around the edges of the mask, so all the air that the animal breathes out gets analysed. Suppose that in an experiment the flow meter records a flow rate of 10 litres of air per minute. Suppose also that the fresh air contains 20.95 per cent oxygen by volume (the normal proportion) and that the analyser shows that the air passing through it contains 19.85 per cent oxygen by volume. Then the oxygen being used per minute is 1.10 per cent of 10 litres or 110 cm^3. Notice that it does not matter for this calculation that much of the air has passed through the mask without being breathed; what matters is that all the air that has been breathed passes through the flow meter and analyser. For small animals that cannot conveniently be fitted with a mask, a transparent hood is fixed over the animal, leaving just a narrow gap between the edges of the hood and the treadmill surface (Fig. 4.1(b)). Air is drawn from the hood for analysis in the same way as from the masks on the larger animals.

The energy cost of walking or running can be calculated from the oxygen consumption (remember that metabolism using 1 cm^3 oxygen releases 20 joules) but only if the speed is low enough for locomotion to be powered entirely by aerobic metabolism. Section 4.5 will show how sprinting depends on anaerobic metabolism in which glucose is converted to lactate without any oxygen being used. In cases of doubt it is necessary to check that no substantial anaerobic metabolism is happening, by measuring the lactate concentration in the blood.

In a particularly informative set of experiments, Taylor and his student Dan Hoyt measured the oxygen consumption of small ponies on a treadmill. Like other mammals, ponies use different gaits at different speeds; they walk to go slowly, trot at intermediate speeds, and gallop to go fast. Each gait involves a different footfall pattern, as shown in Fig. 4.2. Hoyt and Taylor were able to train the ponies to change gaits on command, so were able to make them trot at speeds at which they would normally have walked and to gallop at speeds at which they would have

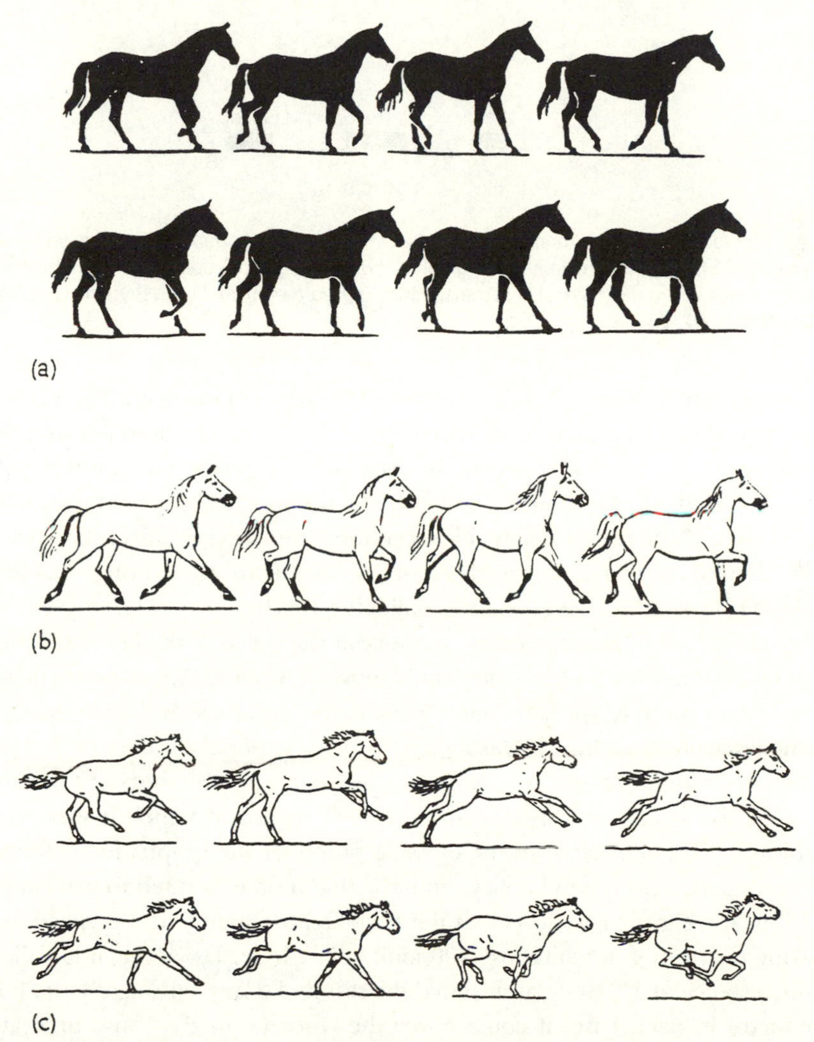

Fig. 4.2 The three principal gaits of horses: (a) walking, (b) trotting, and (c) galloping. In walking, each foot is on the ground for more than half the time, but in the two running gaits (the trot and the gallop) each foot is on the ground for less than half the time. The sequence in which the feet are set down is distinctive for each gait. The canter (not illustrated) is a slow form of gallop. (From P. P. Gambaryan (1974) *How Mammals Run*, Wiley, New York.)

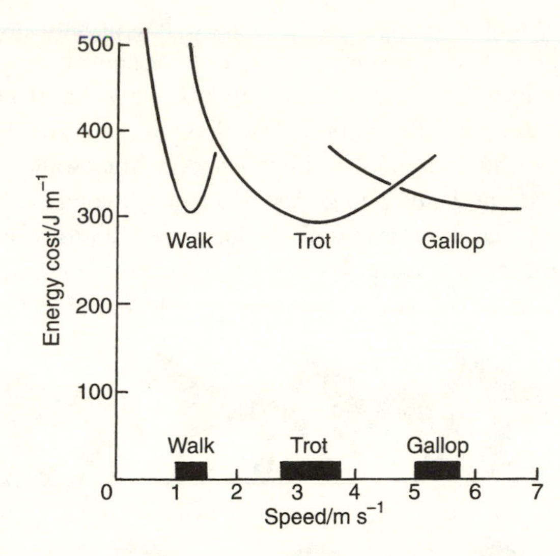

Fig. 4.3 Metabolic energy per unit distance used by 100 kg ponies walking, trotting, and galloping on a treadmill. Bars at the bottom of the graph show the ranges of speeds at which the ponies used each gait when moving freely in their paddock. (Redrawn from D. F. Hoyt and C. R. Taylor (1981) *Nature* **292**, 239–40.)

preferred to trot. Figure 4.3 shows results from the experiment. The vertical axis shows metabolic rate divided by speed, which is the energy used per unit distance travelled. At any particular speed, the metabolic rate depended on the gait being used. At speeds up to 1.7 m s^{-1} walking was the most economical gait; when the ponies were made to trot below this speed they used oxygen faster than when they were allowed to walk. At speeds from 1.7 to 4.6 m s^{-1} trotting was the most economical gait and above 4.6 m s^{-1} galloping was best.

On the treadmill the ponies had to move at the speed of the belt, but when they were set loose in their paddock they could move at whatever speed they would. They were filmed moving spontaneously around the paddock and their speeds were calculated from the films. It was found that they generally walked at 1–1.5 m s^{-1}, trotted at around 3 m s^{-1}, or galloped at 5–6 m s^{-1} (see the bars at the bottom of Fig. 4.3). Each gait was used in the range of speeds at which it was most economical, and speeds close to the crossing points of the graphs for different gaits were avoided. This is good policy. Suppose that a pony wanted to make a journey at 1.7 m s^{-1}. Figure 4.3 shows that if it travelled constantly at this speed, whether walking or trotting, it would use 400 joules per metre. However, if it walked part of the distance at 1.2 m s^{-1} and trotted the rest at 3.2 m s^{-1} (using about 300 joules per metre in each case) it could cover the distance in the same time at lower energy cost.

Figure 4.3 shows that when 100 kg ponies move at their preferred speeds they use about 300 joules for every metre they travel, irrespective of whether they are walking, trotting, or galloping. They use energy faster when they travel fast, but

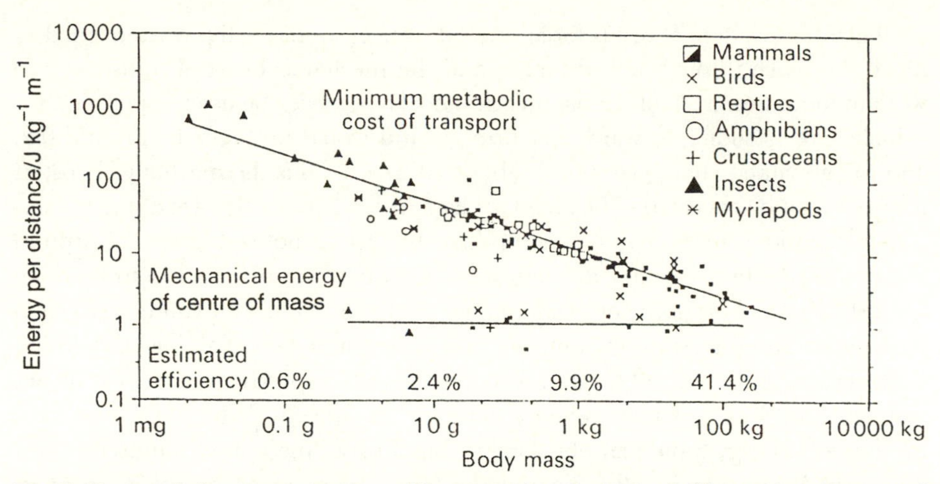

Fig. 4.4 The metabolic energy costs for walking or running of animals of different sizes. Net cost of transport, shown on the vertical axis, is the energy needed (in excess of what would be used when standing still) to move unit mass of animal over unit distance. It is plotted against body mass (both on logarithmic scales). The lower line shows the mechanical cost of transport. (From R. J. Full (1989) in W. Wieser and E. Gnaiger (eds) *Energy Transformations in Cells and Organisms* pp. 175–82, Thieme, Stuttgart.)

the energy cost per unit distance is about the same for all three gaits. Similarly for most other animals that have been investigated, the energy cost per unit distance of legged locomotion is about the same over a wide range of speeds. This makes possible the comparison of animals of different sizes that is shown in Fig. 4.4.

The 'metabolic cost of transport' shown in this graph is the energy needed to move a kilogram of animal a distance of one metre. An ant needs less energy to move a metre than a pony, but not in proportion to its mass; the cost of transport is about 1000 J kg^{-1} m^{-1} for ants but only about 3 J kg^{-1} m^{-1} for ponies. Over the range of sizes investigated, cost of transport is about proportional to (body mass)$^{-0.32}$.

Similar-sized animals generally have fairly similar costs of transport, even if their anatomies are very different. Figure 4.4 includes data for mammals, birds, lizards, salamanders, insects, and crabs. Other experiments have shown that snakes, which crawl on their bellies instead of running on legs, have similar costs of transport to mammals and lizards of equal mass. However, there is quite a lot of scatter of points on either side of the line, showing different costs of transport for animals of equal mass. For example, three of the points for birds are well above the line. These points are for penguins, which waddle in an ungainly way on land.

Why do different-sized animals have different costs of transport? Is this what we should expect? We will try to answer the question by estimating the work that the muscles have to do. This is simplest for snakes. A snake of mass m presses on the ground with a force mg, where g is the gravitational acceleration. Its movement over the ground is resisted by sliding friction with a force μmg, where μ is the coefficient of friction. Work is force times distance, so μmg is the work needed to move the snake through unit distance, and the mechanical cost of transport (the work to move unit mass of animal unit distance) is μg. The coefficient of friction for a snake's body sliding forward over firm ground would probably be around 0.2, and the gravitational acceleration is about 10 m s^{-2}. Thus the mechanical cost of transport can be expected to be about 2 J kg^{-1} m^{-1}, whatever the size of the snake.

Now consider an animal moving on legs. Its feet do not slide over the ground but are lifted whenever they are moved. Consequently, no work has to be done against friction. This does not mean that no energy is needed for running. In each walking or running step an animal's body rises and falls, gaining and losing potential energy (energy of height). Also in each step it speeds up and slows down, gaining and losing kinetic energy (energy of speed). Whenever the total mechanical energy (potential plus kinetic) has to rise, the animal's muscles must do work. Whenever it falls, the muscles must function as brakes to convert mechanical energy to heat, which is lost. Thus the work is discarded almost as soon as it has been done, but the work is still necessary.

This work has been measured for many animals, for example kangaroos. These hop on their hind legs, moving the left and right legs together, but the basic principles of their locomotion are the same as for humans and animals that move left and right legs alternately.

Alexandra Vernon and I used a force plate to measure the forces that hopping kangaroos exert on the ground. This is a force-sensitive plate set into the floor. When the animal's feet landed on it, it gave electrical outputs representing the magnitude and direction of the force they exerted and the coordinates of the point on the platform at which the force was centred. This enabled us to superimpose the forces on cine-films taken simultaneously (Fig. 4.5). As the figure shows, the force was always more or less in line with the legs. In (a), just after the feet touch down, they are pushing downwards and forwards, slowing the animal down. In

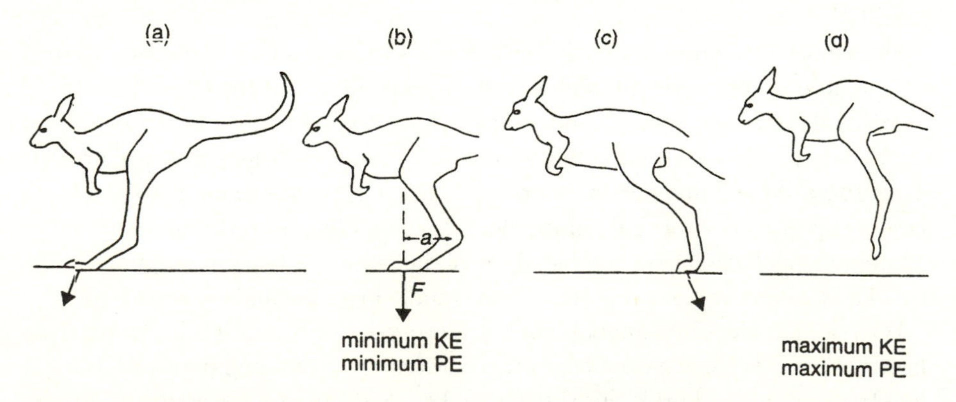

Fig. 4.5 Outlines traced from a film of a kangaroo hopping. The arrows represent the forces that the feet exert on the ground; KE = kinetic energy and PE = potential energy. (From R. McN. Alexander (1990) *The Chordates*, Cambridge University Press.)

(c), shortly before the feet are lifted, they are pushing down and back, speeding the animal up. Thus the kangaroo is travelling most slowly at stage (b) and fastest when its feet are off the ground, at stage (d). Also, because it is progressing in a series of leaps the animal is lowest at stage (b) and highest at stage (d). Its total (kinetic plus potential) energy is lowest at stage (b) and highest at stage (d).

For some calculations, the fluctuations of speed and height were measured from films and used to calculate the energy changes. In other cases the energy changes were obtained more conveniently and accurately, but indirectly, by calculating them from the force record. At higher speeds, more work was needed for each stride but the strides were longer, and the mechanical cost of transport was about 2.5 J kg^{-1} m^{-1} both for slow and for fast hopping.

Similar measurements and calculations have been made for running by other animals; results are shown by the lower line in Fig. 4.4. For most of the legged animals that have been investigated, over a wide range of sizes, the mechanical cost of transport is about one joule per kilogram metre (the long hops of kangaroos need a lot of work).

Here we have a paradox. The metabolic cost of transport is lower for larger animals but the mechanical cost of transport is about the same for animals of all sizes. Small animals are less efficient, needing more metabolic energy to do the same amount of work. For example a 10 g mouse has a metabolic cost of transport of about 30 J kg^{-1} m^{-1} and a mechanical cost of transport of 1 J kg^{-1} m^{-1}, so it does the work of running with efficiency $1/30 = 0.03$. A 70 kg human has costs of transport of about 4 and 2 J kg^{-1} m^{-1} respectively for running, giving an efficiency of about 0.5. However, we know from experiments with muscles taken from the legs of freshly killed mice that they can do work with efficiencies up to 0.25, and measurements of oxygen consumption of people on bicycle ergometers show that human muscles also work with efficiencies of, at best, 0.25. Human and mouse muscles are capable of similar efficiencies but the mouse ones perform much less well in locomotion.

We have two puzzles here—the low efficiency of running of small animals and the high efficiency of humans and other large mammals. The low efficiency of small mammals is an unsolved puzzle. The high efficiencies of large ones, however, can be explained: the work required of their muscles may be less than the (potential plus kinetic) energy changes, because large running animals save energy by the principle of the bouncing ball.

When a rubber ball hits the ground it loses kinetic energy, but this energy is not permanently lost. Most of it is stored briefly in the ball as elastic strain energy, and returned by an elastic recoil to power the next bounce. Some of the kinetic and potential energy lost at each footfall of a running animal is similarly stored as elastic strain energy and returned in an elastic recoil to help power the next step.

The principal elastic structures are tendons. These can stretch elastically by up to 10 per cent of their length without breaking. In their elastic recoil they return 93 per cent of the work done stretching them. At stage (b) in Fig. 4.5 the force on the feet is greatest, the leg tendons are most stretched, and the stored strain energy is greatest. This is the stage at which potential and kinetic energy are lowest. On the other hand at stage (d), when the feet are off the ground, potential and kinetic energy are high but the tendons are relatively slack and hardly any strain energy is stored. Elastic strain energy rises as potential and kinetic energy fall, so the bouncing ball principle reduces the work that the muscles have to do.

To estimate the energy savings we need to know the forces that act on the tendons, and how much the tendons are stretched. The forces were first calculated from force-plate records and films. In Fig. 4.5(b), the force F acts at a distance a from the ankle, so muscles must be exerting a moment aF about the joint. The muscles principally concerned are the gastrocnemius and plantaris (Fig. 4.6), and the forces in them can be calculated. More direct measurements are now possible, using devices called buckle transducers. These are attached to the tendons, a procedure that involves minor surgery. Wires carry the electrical output of the transducers to a radio transmitter attached to the animal's back, which transmits the information to recording equipment. Thus records have been made of forces in leg tendons of wallabies (small kangaroos) as they hopped. Afterwards, the animals were killed and the tendons tested, to find out how much the measured forces stretched them. The forces acting in fairly fast hopping, at 6 m s^{-1}, were found to stretch the tendons by about 3.5 per cent. The energy stored in the stretched tendons amounted to some 40 per cent of the (potential plus kinetic) energy that the muscles absorbed and returned each time the feet were on the ground. The tendons, acting as springs, reduced the work that the leg muscles had to do by about 40 per cent.

The equivalent experiment cannot be done on humans. Even if someone were to submit to the surgery involved, he or she could not be killed afterwards for tests on the elasticity of the tendons. However, the force on the Achilles tendon has been calculated from force-plate records of human running. Hence, knowing the dimensions of the Achilles tendon and the elastic properties of tendons in general,

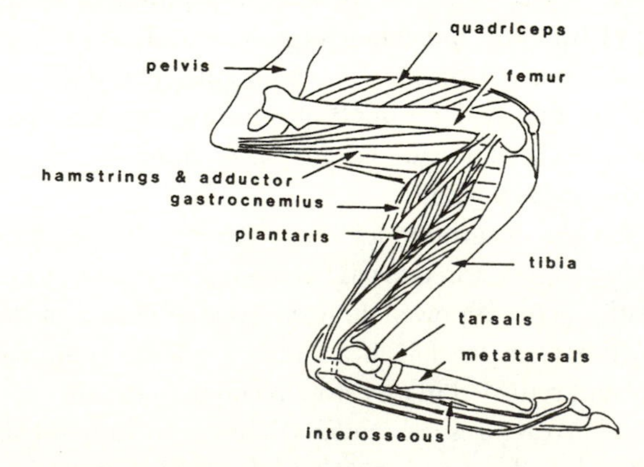

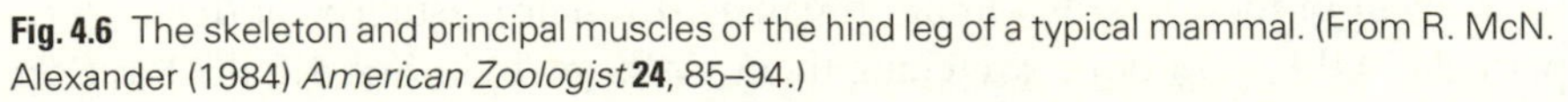
Fig. 4.6 The skeleton and principal muscles of the hind leg of a typical mammal. (From R. McN. Alexander (1984) *American Zoologist* **24**, 85–94.)

the degree of stretching and the stored energy have been estimated. As well as the tendon, the arch of the foot serves as an energy-saving spring; my colleagues and I have measured its elastic properties in experiments with feet that had been amputated for medical reasons. Our conclusion was that, in running at middle-distance speeds, the spring-like properties of the Achilles tendon and of the arch of the foot between them halve the work that the leg muscles have to do. The muscles do not have to work with 50 per cent efficiency, as seemed to be the case a few paragraphs ago, but only 25 per cent. We already know from experiments with bicycle ergometers that this efficiency is possible.

There is a part of the work of running that we have not discussed so far (though account was taken of it in the efficiency calculations). In addition to the kinetic energy changes due to acceleration and deceleration of the body as a whole, there are others due to the movements of parts of the body relative to each other, principally the movements of the legs as they swing backwards and forwards relative to the trunk. The work required to swing the legs is relatively small at low speeds but more important in fast running. My colleagues and I have suggested that the change from trotting to galloping at high speeds may save energy that would otherwise be needed to swing the legs.

Here is how our suggestion went. In walking and trotting, the left leg of each pair moves half a cycle out of phase with the right leg (Fig. 4.2). In galloping, however, the legs of a pair are only slightly out of phase with each other; the two hind legs are set down in rapid succession, and then the two fore legs. The back straightens as the hind legs come off the ground and bends as the fore legs come off the ground. (In dogs and many other mammals, in which the bending is more obvious than in horses, the whole lumbar region of the vertebral column bends; but in horses most of the bending is at the joint between the last lumbar vertebra and the sacrum.) At the stage when the back is most bent (the last outline in Fig. 4.2(c)) the fore legs have been swinging back and are about to swing forwards; the hind legs have been swinging forwards and are about to swing back. All four legs are reversing their direction of swing, so they must lose and regain kinetic energy. However, at this stage the back extensor muscles are active and the sheet of tendon that connects them to the hindquarters is stretched. Kinetic energy taken from the legs may be stored as elastic strain energy in this sheet of tendon and returned by its elastic recoil to help set the legs swinging in the opposite direction. This energy-saving mechanism will be most useful at high speeds, at which the work needed for leg swinging is relatively large. It may explain why, at high speeds, galloping is more economical than trotting (Fig. 4.3).

4.2 **Flight**

The only animals that can fly are birds, bats, and insects. A few other animals such as 'flying' squirrels and the lizard *Draco* have small wings and are quite accomplished gliders, but are incapable of powered flight.

The energy used in flight has been determined for various animals, by measuring oxygen consumption. The principle is the same as for the experiments on running but the details are necessarily different. For measurements of the energy cost of running it was convenient to keep the runner stationary in the laboratory, by having it run on a moving belt. The same effect has been achieved in experiments on flight by training the animal to fly against a jet of air in a wind tunnel. Figure 4.7 shows the wind tunnel used for the earliest measurements, on budgerigars. A powerful fan draws air through the tunnel. The working section where the bird is to fly is separated off by grids, so the bird is in no danger from the fan. If the air flows smoothly and at the same velocity everywhere in the working section, the energy cost for the stationary bird flying in the moving air will be the same as if the air were still and the bird were travelling at the same relative velocity. The tunnel is carefully designed to give smooth, even flow. Another very important point is that the diameter of the tunnel must be much larger than the animal's wing-span, otherwise 'wall effects' will alter the power needed for flight.

It is not too difficult to train birds to fly in wind tunnels. In training sessions for his pioneering experiments, Vance Tucker used to sit on the floor of the working section. Whenever the bird landed, Tucker picked it up and threw it into the air again. The bird soon got the message and Tucker was able to leave the tunnel. The bird wore a mask from which the air was drawn away for analysis, in the same way as in the experiments on running (Fig. 4.1).

In experiments on bumble-bees, Charles Ellington decided (understandably) that it was impracticable to fit his subjects with masks. Instead, he made a tiny wind tunnel which cycled the same air repeatedly through the working section. He used an exceptionally sensitive oxygen analyser to measure the declining oxygen concentration in the air.

Results from experiments with birds and bats are shown in Fig. 4.8(a). In some cases (for example starlings) the metabolic rate is almost constant over the whole range of flying speeds. It is also nearly constant in bumble-bees, which use about

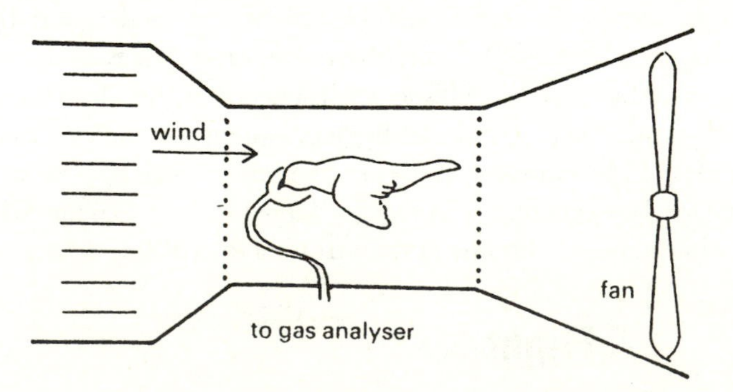

Fig. 4.7 A diagram of the wind tunnel used by Dr Vance Tucker for measuring the oxygen consumption of birds in flight. The bird is drawn unrealistically large. (From R. McN. Alexander (1982) *Locomotion of Animals*, Blackie, Glasgow.)

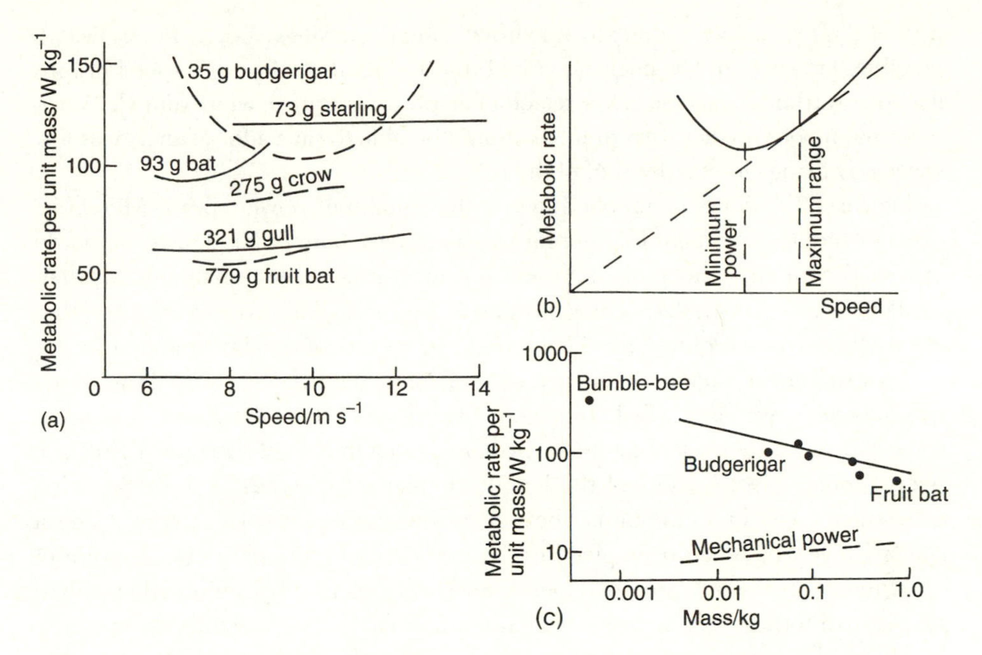

Fig. 4.8 The energy cost of flight. (a) Metabolic rate per unit body mass plotted against speed, for various birds and bats. (b) How a minimum power speed and a maximum range speed can be found if the graph of metabolic rate against speed is U-shaped. (c) A graph of minimum metabolic rate per unit body mass in flight against body mass. The points refer to the same species as (a), together with a bumble-bee. The continuous regression line refers to a larger sample of birds and bats. In addition, the broken line shows estimates of the mechanical power, per unit body mass, required for flight by birds and bats. ((a) redrawn from R. McN. Alexander (1982) *Locomotion of Animals*, Blackie, Glasgow. The lines in (c) were calculated from equations in J. M. V. Rayner (1995) *Israel Journal of Zoology* **41**, 321–342. The point for the bumble-bee is from C. P. Ellington *et al.* (1990) *Nature* **347**, 472–3.)

350 W kg^{-1} at all speeds. Some other birds, notably budgerigars, have been found to have U-shaped graphs of metabolic rate against speed. This is what would be expected by analogy with aircraft, which similarly have U-shaped graphs of power requirement against speed. Both for aeroplanes and for budgerigars there is a minimum power speed at which the power needed for flight is least.

A pilot who wants to remain airborne for as long as possible without running out of fuel should fly at the minimum power speed. However, if she or he wants to fly as far as possible it is better to fly a little faster, and arrive sooner. Figure 4.8(b) shows how the best speed in this case, the maximum range speed, can be found. The slope of a line from the origin to a point on the curve equals (metabolic rate/speed) or (energy cost/distance). The point for which this slope is least is the one for which the line is a tangent, and the tangent is shown on the graph. This indicates the speed that minimizes (energy cost/distance) and so gives maximum range, for birds such as budgerigars that have U-shaped graphs of metabolic rate against speed. No such tangent can be drawn to graphs like the one for the starling

(Fig. 4.8(a)) so it seems that, to maximize range, starlings should fly as fast as possible. However, it has been suggested that the flatness of power–speed graphs like the starling's may be an artefact of experiments in a wind tunnel. Aerodynamic theory predicts that animals should be able to save a lot of energy at low speeds by flying close to the roof of the tunnel.

Figure 4.8(c) shows metabolic rates at the minimum power speed. Metabolic rate per unit mass in flight is larger for small animals than for large ones. Similarly at rest, the minimal metabolic rate per unit mass is larger for small animals than for large ones (Section 2.4). The general rule for birds of all sizes seems to be that the metabolic rate required for flight is 7–10 times the minimal metabolic rate.

To explain the metabolic energy cost of flight we need to know how much mechanical power is needed. In recent experiments, Ken Dial and colleagues have measured the work done by the principal wing muscle of a magpie flying in a wind tunnel. They calculated the length changes of the muscle from the wing movements seen in a cine-film. They measured the forces it exerted by using a strain gauge which had been glued to the humerus, close to the muscle's point of attachment, in a simple surgical operation. Forces exerted by the muscle resulted in slight distortions of the bone, which were detected by the strain gauge. The mechanical power output of the muscle, at each stage of the wing beat cycle, was the force multiplied by the rate of shortening. They found that the total power (both wings) averaged over a wing beat cycle was 21 W kg^{-1} when the bird was hovering; about 9 W kg^{-1} over a wide range of speeds, from 4 to 12 m s^{-1}; and about 12 W kg^{-1} at 14 m s^{-1}, the fastest it could be persuaded to fly. These mechanical powers are well below the metabolic powers required for flight by a 170 g bird (Fig. 4.8(a)), but that is what we should expect since muscles generally work with efficiencies of 0.25 or less.

We can get a more fundamental understanding of the energy cost of flight by examining the aerodynamic mechanisms involved. The basic principle is that air must be driven downwards. If the animal is able to push down on the air with a force equal to its body weight, the air will support it.

The air driven downwards by flying birds and bats has been made visible in experiments at Bristol University, by a team led initially by Professor Colin Pennycuick. The room was filled with a cloud of tiny soap bubbles, of about 3 mm diameter. Ordinary soap bubbles sink in air but these ones were filled with a helium–air mixture adjusted to give the complete bubbles the same density as air; they had very little tendency either to rise or to sink in still air. The animal flew through the cloud of bubbles, disturbing the air, and the movements of the bubbles showed how the air was moving. The pattern of air flow in the animal's wake was revealed by taking stereoscopic multiple-flash photographs.

To understand the patterns seen in the photographs, we need to understand vortices. These are cylinders or rings of air set rotating by a moving body of air, in the same way as rollers rotate when a crate is pushed over them. Whenever a jet or puff of air moves through still air, vortices form around it. A smoke ring is a ring-shaped vortex with smoke trapped in it, formed around a puff of air from a smoker's mouth.

Experiments on the bat *Nyctalus*, flying through the cloud of bubbles, were particularly revealing. When it flew slowly, at about 3 metres per second, each downstroke of its wings drove a puff of air downwards and formed a vortex ring, but the upstrokes produced no perceptible air movement. Thus flight was powered entirely by the downstroke. Figure 4.9(a) shows a series of vortex rings produced by successive downstrokes.

When the bat flew fast, at about 8 metres per second, air was driven downwards continuously, not in a series of puffs. The upstroke as well as the downstroke did useful work. Throughout the wing beat cycle, the air immediately behind the wings was moving downwards, and continuous vortices extend back from the wing tips (Fig. 4.9(b)). These resemble the wing tip vortices formed by aeroplanes, which in some conditions become visible as vapour trails. However, they are not straight and parallel like vapour trails, but undulate with the wing beat.

The two patterns of flight, forming vortex rings or continuous vortices, are distinct gaits used at different speeds, like walking and running. Most birds also use the vortex ring gait at low speeds and the continuous vortex gait to fly fast. I used a bat to introduce the gaits only because it was the first animal to be shown to use both.

Most insects, and hummingbirds, fly differently. The principle is that of helicopters although the wings, of course, flap backwards and forwards instead of whirling continuously in one direction like a helicopter rotor. When a helicopter

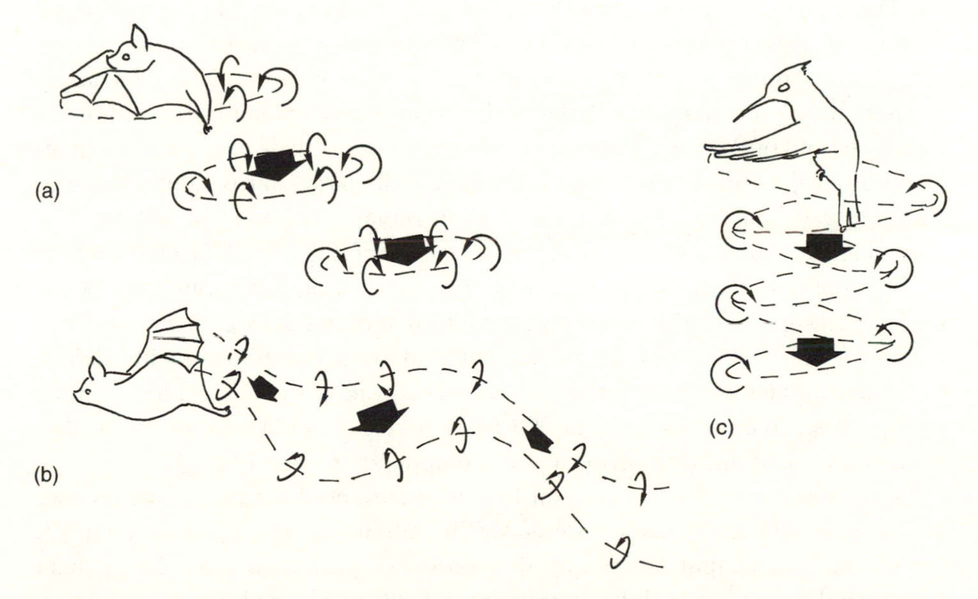

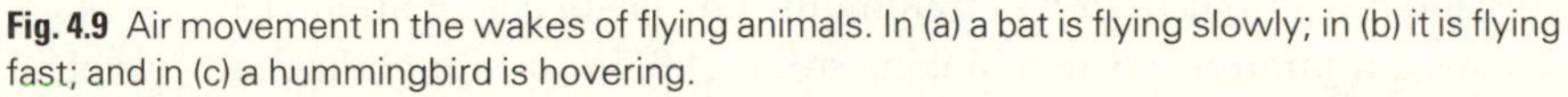
Fig. 4.9 Air movement in the wakes of flying animals. In (a) a bat is flying slowly; in (b) it is flying fast; and in (c) a hummingbird is hovering.

hovers it keeps its rotor horizontal, to drive air vertically downwards. Similarly a hovering hummingbird, bee, or moth beats its wings in a horizontal plane, producing a stack of linked vortex rings (Fig. 4.9(c)). To fly forward a pilot tilts the helicopter so as to blow air downward and backward, providing thrust as well as weight support. Insects fly forward by beating their wings in a suitably tilted plane.

Let us see how the aerodynamics helps us to understand mechanical power requirements. We can use helicopter theory to get rough estimates of power for hovering. The power per unit mass needed for hovering, by a helicopter of mass m with a rotor of radius r, is $(mg^3/2\rho\pi r^2)^{0.5}$, where g is the gravitational acceleration (about 10 m s^{-2}) and ρ is the density of air (1.2 kg m^{-3}). For a 5 g hummingbird with 60 mm wings this is 14 W kg^{-1}. For a 0.2 g bumble-bee with 14 mm wings it is 12 W kg^{-1}. These are minimum possible values. They take account of the kinetic energy that the wings must give to the air that is driven downwards, but nothing else.

Notice that the minimum possible power per unit mass is proportional to $(m/r^2)^{0.5}$. If insects of different sizes had exactly the same proportions (i.e. were geometrically similar), wing length r would be proportional to the cube root of mass m, and this power per unit mass would be proportional to $m^{1/6}$; it would be greater for large insects than for small ones. (Our calculated values for humming-birds and bees are about the same despite the difference in size because humming-birds have relatively long wings and bees have relatively short ones.)

The helicopter analogy can be applied also to ordinary birds (i.e. not hummingbirds) when they are hovering or flying slowly, using the vortex ring gait. For Dial's 170 g magpies it gives a power estimate of 18 W kg^{-1}, in fair agreement with the observed power of 21 W kg^{-1}. The conclusion from helicopter theory, that power requirements per unit mass can be expected to be larger for larger animals, helps to explain why large birds cannot hover. Small birds such as tits hover well and often do so, for example when arriving at their nest. Dial's magpies could hover for only 1–4 seconds, and substantially larger birds cannot hover at all. Note that hovering means flying so as to remain stationary in still air. The 'hovering' of kestrels (*Falco tinnunculus*) is not true hovering, as will be explained.

The continuous vortex gait that birds use when flying fast is quite unlike the flight of helicopters, and is better compared to that of fixed-wing aircraft. We can make a rough estimate of the power required by a comparison with gliders. Consider a glider of mass m that is losing height at a rate v_{sink}. The potential energy it loses in unit time is mgv_{sink}. To avoid sinking it would need an engine that would do work at this rate, giving a power output *per unit mass* of gv_{sink}.

Good man-made gliders in still air lose height at about 0.5 metres per second. Small model gliders fly much more slowly, but also lose height at about 0.5 m s^{-1}. No gliding animal that has been tested performs quite as well as this (a small vulture and a fruit bat both had minimum sinking speeds of about 1 m s^{-1}) but if we assume a minimum-possible sinking speed of 0.5 m s^{-1} for gliding animals of all sizes we can estimate the minimum power requirement for level flight (using the expression above) as 5 W kg^{-1}.

More complicated and accurate calculations for birds and bats gave the values shown by the lower line in Fig. 4.8(c). The lines for metabolic and mechanical power converge as mass increases, showing that larger animals fly more efficiently. Small (10 g) birds fly with efficiencies around 0.04 and large (1 kg) birds with efficiencies around 0.3. This tendency parallels the one we have already noted for running. As for running, it is difficult to explain.

Many large birds save energy by soaring whenever possible, using natural air movements to keep themselves airborne instead of flapping their wings. There are two principal techniques, slope soaring and thermal soaring.

Slope soaring can be done in places where the wind blows against a hillside or a large wave and is deflected upwards (Fig. 4.10(a)). Gulls and other sea birds can often be seen soaring in front of coastal cliffs, and albatrosses soar along waves. They are gliding without flapping their wings so must sink *relative to the air*, but if the air is rising sufficiently fast they can remain airborne indefinitely.

Soaring albatrosses can travel at an angle to the waves by alternately soaring along a wave and gliding at right angles to the waves. Kinetic energy built up in the soaring phase carries them through the glide to the next wave. Measurements of

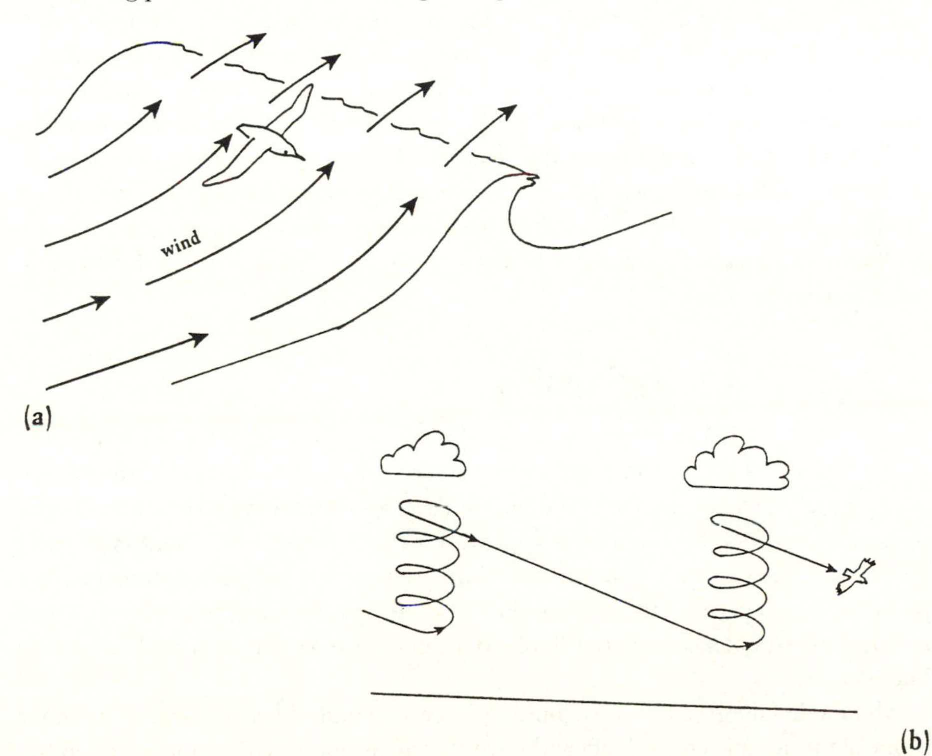

Fig. 4.10 (a) Slope soaring and (b) thermal soaring. (From R. McN. Alexander (1989) *Dynamics of Dinosaurs and other Extinct Giants*, Columbia University Press.)

the heartbeat frequencies of free-flying albatrosses (to be described in Section 7.7) indicate metabolic rates during soaring of only 5 W kg^{-1}, far less than would be needed for powered flight by a similar-sized (3.6 kg) bird (see Fig. 4.8(c)). This is only 2.2 times the estimated minimum metabolic rate.

Kestrels hunting for small rodents commonly soar over hillsides, facing directly into the wind. By adjusting their gliding speed to match the wind speed they can keep themselves stationary. This is the behaviour that is often misleadingly described as 'hovering'.

The other soaring technique used by many large birds is thermal soaring. Land heated by the sun gets hotter than the air. The hot ground in turn heats the air close above it, so that there is hot air close to the ground and cooler, denser air higher up. This is an unstable situation and the hot air rises in columns known as thermals. The pattern of air movement is like the movement of water in a heated saucepan that has not yet come to the boil. Cumulus clouds form at the tops of thermals which helps glider pilots (and presumably birds) to find them.

Vultures, storks, and other birds (and human glider pilots) soar in thermals. They are gliding, sinking relative to the air, but if the air is rising fast enough the bird will gain height relative to the ground. The usual technique is to circle for a while in one thermal, gaining height, and then glide to another (Fig. 4.10(b)). I have no information about the metabolic rates of birds soaring in thermals but it seems likely that this kind of soaring is as economical of energy as slope soaring.

Vultures in the Serengeti cannot soar until mid morning because it takes a few hours for the morning sun to heat up the ground. Once the thermals are strong enough they take off and soar all day, except when they land to feed on carcasses. Professor Colin Pennycuick used a powered glider to follow and observe them. On one occasion he followed a vulture that was returning to its nest after feeding 75 kilometres away. It covered this distance in 96 minutes, entirely by soaring, using just six thermals.

4.3 **Swimming**

There are many kinds of swimming animals that swim in many different ways. This section is largely about the teleost fishes and the squids, two very plentiful groups that include many of the most impressive swimmers. Ducks, penguins, and dolphins will also appear. I have omitted very small swimmers (protozoans, planktonic crustaceans, and invertebrate larvae) although they are so plentiful, partly because their swimming has been less thoroughly studied but also because the cost of swimming probably does not figure very largely in their energy budgets.

Metabolic energy costs of swimming have generally been measured in water tunnels, the aquatic equivalents of the wind tunnels that have been so useful in the study of flight. Figure 4.11(a) shows a water tunnel designed for experiments on fish. A powerful pump drives the water repeatedly around the circuit of wide-bore pipes. The same precautions are taken as in wind tunnels to make the flow in the working section as smooth and uniform as possible. The fish swims against the current so as to remain stationary relative to the walls of the tunnel. It is quite easy to train fish to do this, and this behaviour is natural for fish that live in rivers and swim against the current to avoid being washed downstream. If necessary, the grid behind the fish can be electrified, so that the fish gets a very mild electric shock if it allows itself to drift backwards. As the fish swims it uses oxygen and the dissolved oxygen concentration in the water falls. The declining oxygen concentration is measured by means of an oxygen electrode, so if the volume of water in the system is known the fish's metabolic rate can be calculated. The water is aerated or replaced before the oxygen concentration has fallen too low.

Figure 4.11(b) shows how the equipment is modified for experiments with ducks and other air-breathing swimmers. This is not an enclosed water tunnel but a flume, with the surface of the water in the working section exposed to the air. A transparent hood is suspended just above the water surface and air is drawn from the hood for analysis, exactly as in experiments on running (Fig. 4.1(b)).

Figure 4.12(a) shows the metabolic rate of a fish swimming at different speeds. The faster it goes, the greater the metabolic rate. The graph is far from straight, but curves very sharply upwards. The broken line is a tangent from the origin. The speed at which it touches the curve is the maximum range speed, at which the fish could swim furthest without running out of food reserves. (The argument proving this is the same as for flight, Fig. 4.8(b).)

Figure 4.12(c) shows that big fish tend to have higher maximum range speeds than small ones, as might be expected. Maximum range speed is about proportional to (body mass)$^{0.17}$, or to the square root of body length; as a general rule, if one fish is four times as long as another it should swim twice as fast. Large fish also, unsurprisingly, need more energy to power their swimming. The slope of the line in Fig. 4.12(b) shows that the power required for swimming at the maximum range speed is about proportional to (body mass)$^{0.8}$. The exponent (0.8) is close to the exponent for the relationship between minimum metabolic rate and body

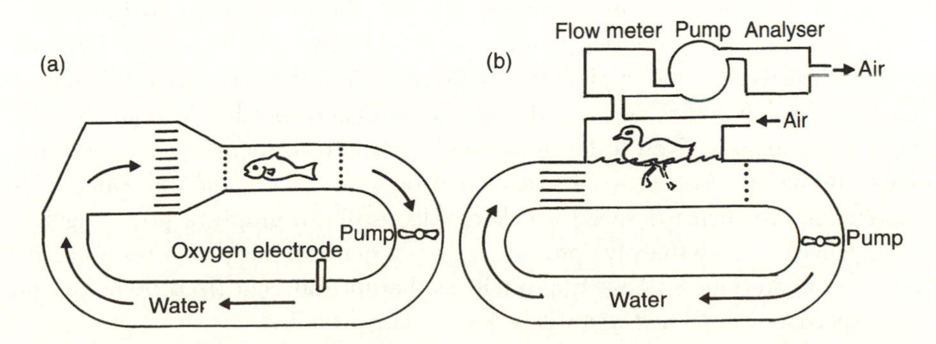

Fig. 4.11 Equipment used for measuring the metabolic energy cost of swimming for (a) fishes and (b) ducks.

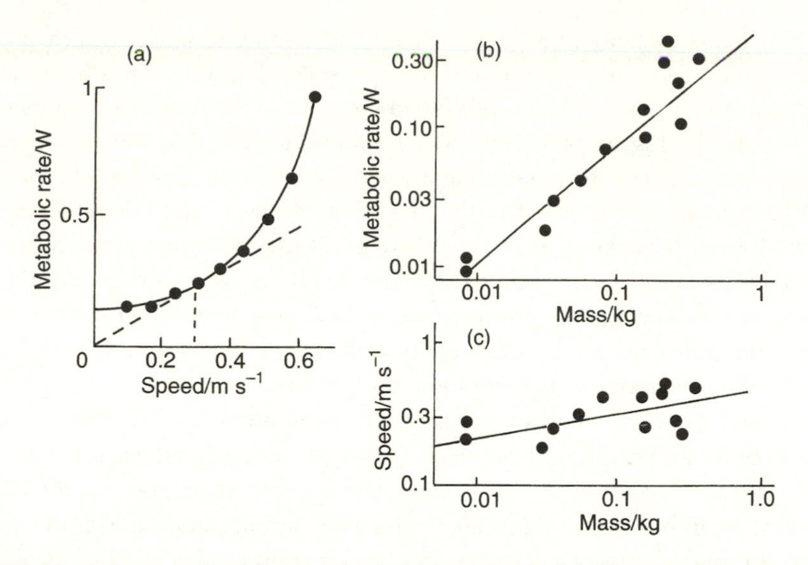

Fig. 4.12 Metabolic rates of swimming animals. (a) A graph of metabolic rate against speed for 0.26 kg trout with a tangent (broken line) indicating the maximum range speed. (b) Metabolic rate at the maximum range speed, plotted against body mass. (c) Maximum range speed, also plotted against body mass. The lines in (b) and (c) were calculated from data for a wider range of sizes of fish than is shown here, including larvae as well as mature fish. (Data in (a) are from P. W. Webb (1971) *Journal of Experimental Biology* **55**, 521–40. Data in (b) and (c) are from J. J. Videler (1993) *Fish Swimming*, Chapman & Hall, London.)

mass. As a very rough general rule, a fish of any size swimming at its maximum range speed will be metabolizing at about double its minimal rate. Larger fish travel faster for a comparable energy cost.

Now we ask, how much work is needed for swimming? The mechanical power needed to propel a rigid body of volume V at speed v through water of density ρ is $0.5\rho V^{0.67}v^3C_{dv}$. C_{dv} is a factor known as the drag coefficient based on volume, which depends on the shape of the body. (This expression is not appropriate for very small or slow swimmers such as protozoans.) If the body has the same density as the water, its mass m equals ρV and the expression can be written $0.5\rho^{0.33}m^{0.67}v^3C_{dv}$. The density of water is 1000 kg m^{-3} (the difference between fresh and sea water is too small to matter for this calculation). For well streamlined (torpedo-shaped) bodies over a wide range of sizes and speeds, C_{dv} is about 0.02. Thus if the mass is given in kilograms and the speed in metres per second, the power needed to propel a well streamlined body is about $0.1m^{0.67}v^3$ watts. This expression, in which the speed is cubed, tells us that a graph of power against speed will curve very sharply upwards, as graphs of metabolic power for fishes are observed to do (Fig. 4.12(a)). It also tells us that for different sized bodies at the same speed, power *per unit mass* will be less for larger bodies.

Now consider a particular fish, a trout 0.3 metres long with a mass of 0.26 kg swimming fast, at 0.6 m s^{-1} (Fig. 4.12(a)). The expression we have just derived tells us that if it were a rigid body, the power needed to propel it would be 0.01 watts. A different calculation gives a much higher power requirement. By analysing film of the fish swimming, considering the momentum that the tail gives to the water as it pushes it to one side or the other, it is possible to calculate the forces on the tail and the work that its muscles must be doing. This method gives a power requirement for our fish of 0.07 watts. The reason for this being so much larger than the rigid-body estimate seems to be that the undulating movements of the swimming fish increase the drag (the resistance to movement through the water) by producing a steeper gradient of velocity in the boundary layer (the thin layer of water immediately against the fish's body). The metabolic power when swimming at 0.6 m s^{-1} is 0.5 watts above the minimal rate (Fig. 4.12(a)). If this is the metabolic power used for swimming, the muscles are working with efficiency $0.07/0.5 = 0.14$. This is within the range of efficiencies we have already calculated for running and flight.

Many fish do not beat their tails when swimming slowly, but propel themselves by movements of other fins. For example, 16 cm bluegill sunfish (*Lepomis*) swim with their pectoral fins at speeds up to 0.2 m s^{-1} and by tail beating above that speed. Thus they seem to avoid the drag-increasing effect of body undulation until they are travelling too fast for their fins alone to propel them.

Squids swim quite differently from fishes. Instead of pushing on the water with a tail they squirt jets of water from their mantle cavities. Many of them are beautifully streamlined (Fig. 4.13(b)) and you might suppose that the energy cost of their swimming would be low. However, measurements of the oxygen consumption of

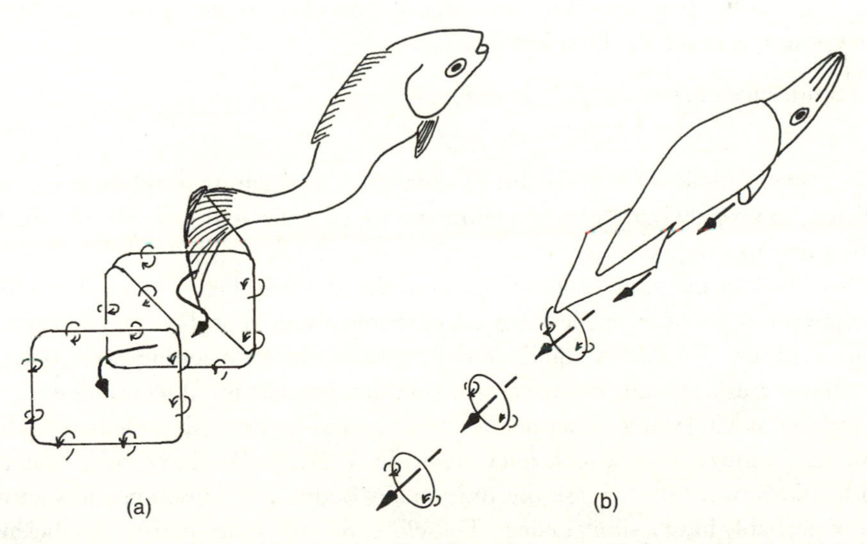

Fig. 4.13 (a) A tuna and (b) a squid swimming. The fish produces a continuous wake, a wavy jet of water threaded through a chain of linked vortex rings, one ring for each tail beat. The squid ejects a series of separate puffs of water, each with its individual vortex ring. The wakes are drawn as they would appear when the animals were accelerating.

squid swimming in a water tunnel showed that they used much more energy than similar-sized trout swimming at the same speed.

There is a basic hydrodynamic principle that seems to explain this. When an animal swims by pushing water backwards it not only has to do work against drag on its body, it has to do additional work, giving kinetic energy to the water that it pushes backwards. To calculate this, we first need to know how much water is accelerated, and to what velocity.

Newton's second law of motion tells us that force equals mass times acceleration. Another way of saying the same thing is that force equals rate of change of (mass × velocity), that is rate of change of momentum. Suppose that in unit time an animal accelerates a mass m_{jet} of water to velocity v_{jet}. It is giving momentum to the water at a rate $m_{jet}v_{jet}$. The force it is exerting on the water equals this rate of change of momentum and must balance the drag D on the body

$$D = m_{jet}v_{jet}$$

The power needed to overcome this drag is the drag multiplied by the swimming speed v_{swim}

$$P_{drag} = m_{jet}v_{jet}v_{swim}$$

Additional power is needed to give kinetic energy to the water pushed backward in the wake. In unit time, mass m_{jet} is being accelerated to velocity v_{jet}, so the rate at which the water is being given kinetic energy is

$$P_{wake} = 0.5m_{jet}v_{jet}^2$$

The ratio of the power used for the essential job of overcoming the drag, to the total power, is called the Froude efficiency:

$$\begin{aligned}\text{Froude efficiency} &= P_{drag}/(P_{drag} + P_{wake}) \\ &= v_{swim}/(v_{swim} + 0.5v_{jet}) \qquad (4.1)\end{aligned}$$

This equation tells us that it is best to make v_{jet} as small as possible. It is more efficient to propel yourself by accelerating a lot of water to a low velocity, than a little water to a high velocity.

Now look at the pattern of flow in the wake of a fish (Fig. 4.13(a)). The tail is pushing on a lot of water, leaving a continuous wake. The Froude efficiency is high, typically 0.7–0.8. A squid, however, expels in each jetting cycle only as much water as its mantle cavity can contain, and its wake is a discontinuous series of puffs of moving water. It accelerates much less water per unit time than the fish and has a lower Froude efficiency, typically 0.3–0.4. We have seen that the undulations of a fish increase the drag on its body. The pulsations of a jetting squid probably have a similar effect. Therefore, the difference in Froude efficiency can be expected to make squid jetting more costly (in terms of energy) than fish swimming.

Another comparison, between 1.1 kg ducks and 1.2 kg penguins, shows different energy costs of different mechanisms of swimming. The measurements were done in a flume like the one shown in Fig. 4.11(b). When both species swam on the surface of the water at 0.7 metres per second, the metabolic rates were 20 watts for the ducks and only 14 watts for the penguins. The ducks used 7 watts and the penguins 8 watts when resting on the surface, so the additional power needed for swimming was 13 watts for the ducks and only 6 watts for the penguins; penguin swimming is much more economical than duck swimming. This may be another example of swimmers with different Froude efficiencies. Ducks propel themselves by paddling, pushing a relatively small mass of water backwards with each foot stroke. Penguins 'fly' through the water, pushing on a large mass of water with each wing stroke. Thus penguins are expected to have higher Froude efficiency. However, they may also have had less drag to overcome.

Swimming at the surface needs more power than underwater swimming because the body has to push a bow wave in front of it, like the bow wave of a boat. When the penguins were persuaded to swim totally submerged (by leaving less air space above the flume), they used less energy; only 11 watts at 0.7 m s^{-1}.

Observations in aquaria show that tunas and typical squids swim unceasingly, day and night. They have to do this (using a lot of energy) because they are denser than water and would sink if they stopped. Most of the tissues of animals are denser than either fresh water (1000 kg m^{-3}) or sea water (1026 kg m^{-3}); for example, the density of muscle is generally about 1060 kg m^{-3}. A complete animal will be denser than the water it lives in unless it has a special buoyancy organ. Being dense may be no problem for swimmers that live near enough to the bottom to rest on it when they have no need to travel. For example, plaice (*Pleuronectes*) and octopus are denser than sea water and spend much of their time on the bottom. However, tunas and typical squids live near the surface of deep water, and resting on the bottom is not an option for them. Might they not save energy by evolving buoyancy organs that would match their densities to the water, so that they had no tendency to sink? Many related animals have done this. Among fishes, most teleosts have gas-filled swim-bladders, and a few sharks contain large quantities of squalene, a low density oil, giving them densities very close to that of the water they live in. Among cephalopods, cuttlefish have gas-filled cuttlebones and the ammoniacal squids (very common in the oceans, within a few hundred metres of the surface) are bloated with huge volumes of ammonium chloride solution, which is less dense than the sodium chloride of sea water; as a result, these cephalopods have about the same density as sea water.

Tunas and typical (non-ammoniacal) squids have fins on either side of the body that function like the wings of aeroplanes, providing upward lift as they move forward through the water and preventing the animal from sinking. Lift cannot be obtained without drag, so energy is needed to move these fins through the water, additional to the energy that would otherwise be needed for swimming. If instead the animals had buoyancy organs, they would avoid the drag on the fins but their bodies would be bigger; swim-bladders generally add about 5 per cent to body volume, and enough squalene to give a shark the same density as sea water adds 25–30 per cent to its volume. If the animal's volume is increased, it will need more energy to swim.

Whether wing-like fins or a buoyancy organ adds less to the cost of swimming depends on how fast the animal swims. The energy cost of wing-like fins does not increase very rapidly with speed because the faster an animal swims, the smaller the fins can be and still supply the required lift. A simple argument predicts that the power requirement will be proportional to speed. The extra power needed due to increased volume, for animals with buoyancy organs, can be expected to be proportional to speed cubed (refer back to the theoretical estimates of the energy cost of swimming). Therefore, we expect buoyancy organs to be the more economical option at low speeds, and wing-like fins at high speeds. To a large extent, animals seem to have evolved the more appropriate option. Among cephalopods, the slow-swimming cuttlefish and the ammoniacal squids (which are believed to be sluggish) have buoyancy organs, and the fast-swimming squids that live near the surfaces of the oceans do not. Among sharks, the basking shark *Cetorhinus*, which swims very slowly filtering plankton from the water, has squalene; and pelagic, predatory sharks get their lift from wing-like fins. Among pelagic teleosts, most have swim-bladders but many of the fast-swimming tunas and their relatives do not.

4.4 **Muscle**

Most animal locomotion is powered by muscles, though some small animals propel themselves by means of cilia. To understand the energetics of locomotion, we need to know about some of the properties of muscle. The information in the next few paragraphs comes from experiments on small muscles or bundles of muscle fibres, dissected from freshly killed animals and made to contract by electrical stimulation.

First, we will look at the results of experiments in which the muscle is stretched or allowed to shorten while being stimulated, and the force it exerts is measured (Fig. 4.14(a)). When it is held at constant length it exerts its isometric force F_{iso}. When allowed to shorten it exerts less force and at its maximum shortening speed v_{max} it exerts no force at all. However, when it is forcibly stretched it exerts higher forces, up to about 1.8 times the isometric force. The power output of the muscle (shown in Fig. 4.14(c)) is the force multiplied by the speed of shortening. It is zero when the speed is zero and also when the speed is v_{max}, which makes the force zero. It has its maximum value at an intermediate speed, generally about $0.3v_{max}$. When the muscle is being forcibly stretched its power output is negative; it is absorbing energy, acting like a brake.

In some experiments, the metabolic energy consumption of muscles has been measured while they are shortening or being forcibly stretched. This has been done on different occasions by measuring oxygen consumption, heat production,

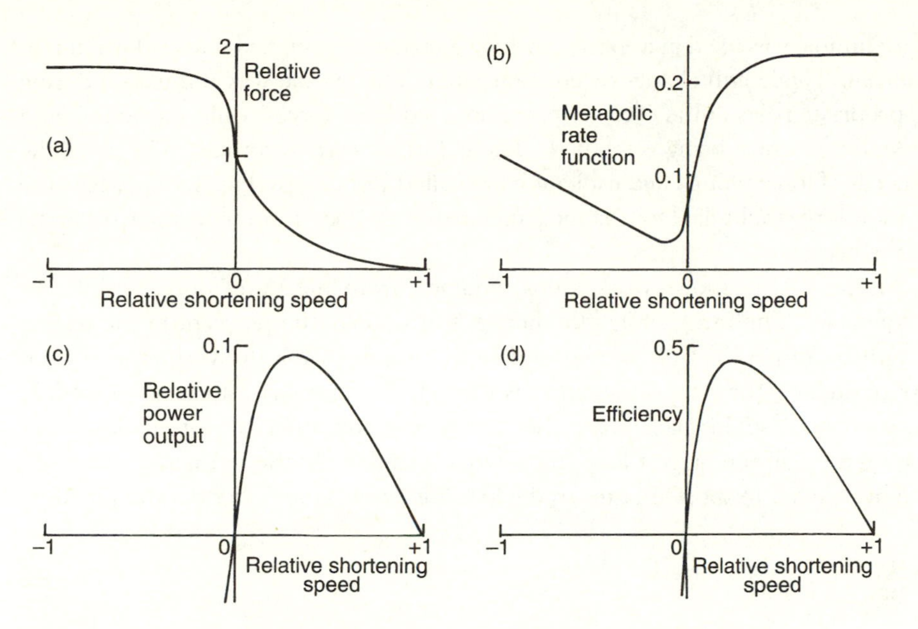

Fig. 4.14 Properties of vertebrate striated muscle, plotted in each case against the rate of shortening. (a) The force exerted by a maximally stimulated muscle; (b) metabolic rate (as ATP consumption); (c) power output; and (d) efficiency. Let the isometric force be F_{iso} and the maximum shortening speed be v_{max}. Then
relative shortening speed means (shortening speed)/v_{max}
relative force means (force)/F_{iso}
relative power output means (power output)/$F_{iso}v_{iso}$
and metabolic rate function means (metabolic rate)/$F_{iso}v_{iso}$.

or ATP consumption. When the length of the stimulated muscle is held constant, its metabolic rate is about $0.07F_{iso}v_{max}$ (Fig. 4.14(b)). When it is allowed to shorten and so do work, its metabolic rate is higher. When it is forcibly stretched at low speeds its metabolic rate is less, but the metabolic rate rises again at faster speeds of stretching.

The efficiency with which the muscle does work is the mechanical power output (Fig. 4.14(c)) divided by the metabolic rate (Fig. 4.14(b)). It is shown in Fig. 4.14(d). It has a maximum value of about 0.45, at a shortening speed of about $0.2v_{max}$. Note that this is the efficiency with which ATP energy is used to do work. The efficiency with which the enthalpy of combustion of foodstuffs is converted to ATP is only about 0.5, reducing the maximum overall efficiency obtainable from muscle to 0.2–0.25, as observed for example in experiments in which the oxygen consumption of athletes is measured while they are pedalling bicycle ergometers.

The experiments on which Fig. 4.14 is based involved single contractions of the muscles, but locomotion generally involves a series of cycles of contraction. In running, legs swing repetitively backwards and forwards. In flight or swimming, wings or fins beat repetitively. Figure 4.15 shows the results of work-loop experiments, which simulate the behaviour of the muscles in a swimming fish. A muscle fibre bundle was fixed in apparatus which alternately stretched it and allowed it to shorten. The length changes were sinusoidal. The bundle was stimulated during appropriate parts of the cycle so that it exerted large forces while shortening and little force while being stretched. These forces were recorded. The resulting records of force plotted against length are called work loops (Fig. 4.15(a)). Because work is force multiplied by distance, the area of the loop represents the work done in the cycle.

Figure 4.15(c) shows work loops obtained from the same fibre at different frequencies. The range of length change, from about 105 per cent of the resting length to 95 per cent of the resting length, was the same in every case. This is approximately the range of lengths over which the fibres are estimated to work in swimming. Also in each case, the timing and duration of stimulation were adjusted to get the biggest loop (most work) possible. As the frequency increases, the muscle has to shorten faster so the forces it exerts are smaller. Consequently it

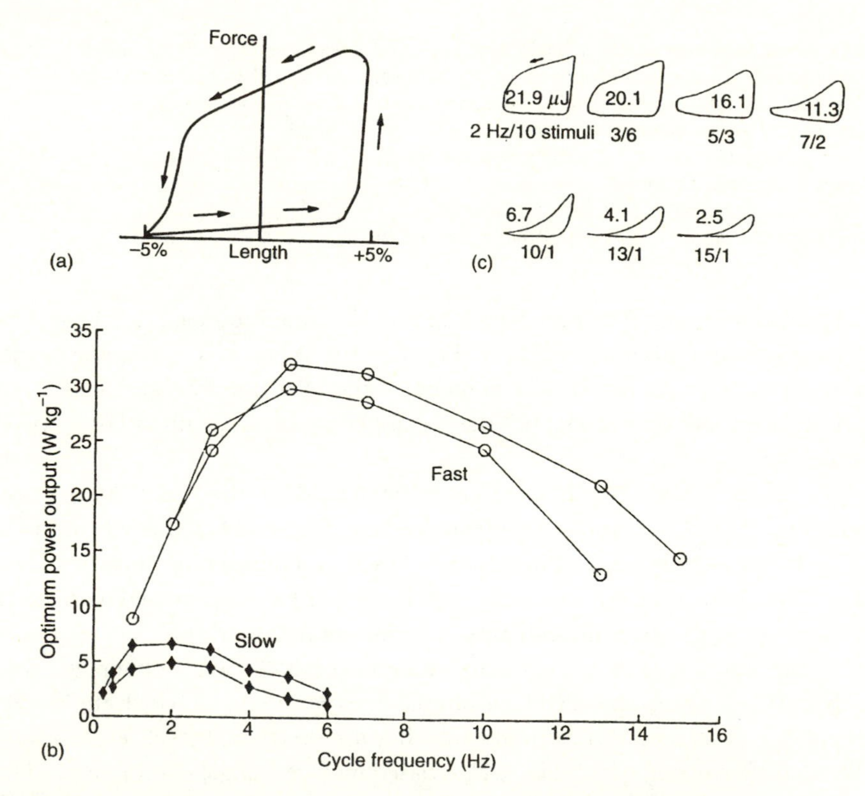

Fig. 4.15 The results of experiments with muscle fibre bundles from a teleost fish, *Myoxocephalus*. (a) A work loop; (b) a graph showing the maximum power output obtainable in a series of work loops, plotted against cycle frequency, for two preparations each of fast muscle fibres and slow ones; and (c) the work loops corresponding to the uppermost curve in (b). (After J. D. Altringham and I. A. Johnston (1990) *Journal of Experimental Biology* **148**, 395–402.)

does less work in each cycle, as frequency increases. However, it gets more cycles in per unit time. At 2 cycles per second it does 21.9 microjoules per cycle, giving a power output of 43.8 microjoules per second (microwatts). At 5 cycles per second it does only 16.1 microjoules per cycle, but this makes the power output much higher, about 80.5 microwatts. At 10 cycles per second the work has fallen to 6.7 microjoules and the power has started to fall; it is 67 microwatts.

Figure 4.15(b) shows power output plotted against frequency for two types of muscle fibre from the swimming muscles of the same fish. In each case, power output is maximal at a particular frequency, about 2 cycles per second for the slow muscle and 6 cycles per second for the fast. This corresponds to a difference in maximum shortening speed, v_{max}, which is much higher for the fast fibres than for the slow ones.

At this stage we need to know more about the two types of muscle and their functions. The main swimming muscles of fishes that power the tail beat make up a large fraction of the body mass (they are the fillets that we eat). The major part of this muscle is white; this is the fast muscle. In addition, most fishes have a strip of red, slow muscle, running along each side of the body just under the skin. Electromyographic experiments have shown how these two kinds of muscle are used. Fine wire electrodes were slipped into the swimming muscles of a fish, one into the white muscle and one into the red. The fish was then set swimming in a water tunnel with wires connecting the electrodes to recording equipment, which recorded the action potentials in the muscles when they were active. These experiments showed that at low swimming speeds, only the slow muscle was active, but at high speeds the fast muscle came into use. For fish of the size and species used for the work-loop experiments (Fig. 4.15), the slow muscle was used when the tail beat frequency was 1–4 cycles per second and the fast fibres in faster swimming with tail beat frequencies of 4–9 cycles per second. These are the ranges of frequencies at which the highest power outputs were obtained in the work-loop experiments.

Other animals as well as fish have fast and slow muscle fibres, and use the faster ones for faster movements, but they are generally mixed together within each muscle. For example, a leg muscle in horses has been found to have slow muscle fibres for which v_{max} averages 0.3 fibre lengths per second and two groups of fast fibres with v_{max} averaging 1.3 and 3.2 lengths per second. The clear anatomical separation of fast and slow fibres is a peculiarity of fish that has made it much easier than it would otherwise have been to find out what the two fibre types do.

In fish swimming, insect flight, and many other repetitive movements, two kinds of forces act—inertial forces to accelerate masses and hydrodynamic forces to overcome drag as they move through water or air. Work is needed to supply kinetic energy whenever a mass is accelerated, and work is also needed to overcome drag whenever a body moves through a fluid. In such repetitive movements, energy may be saved by having springs in the system. These can take up kinetic energy that is lost as the mass decelerates at the end of a stroke, store it as elastic strain energy, and return it in an elastic recoil to supply the kinetic energy for the next stroke. We saw the principle in action in our discussion of running. We will see now that springs can also save energy in swimming and flight.

Figure 4.16 shows two alternative ways in which the springs can be arranged. In each case a pair of muscles has the task of making a plate oscillate up and down in a fluid. In Fig. 4.16(a), a spring is arranged in parallel with the muscles. Any mass mounted on a spring has a natural frequency at which it will vibrate if disturbed. If the stiffness of the spring is chosen to make the natural frequency of the system match the frequency at which the muscles are required to oscillate the plate, the spring will supply the forces needed to overcome the inertia of the plate, and the muscles have only to supply the hydrodynamic forces. This is how scallops such as *Argopecten* are designed. These bivalve molluscs swim by rapidly opening and closing their shells. The two valves of the shell are hinged together, with a piece of rubber-like protein that functions as a spring just inside the hinge. The natural frequency of shell and spring matches the frequency of about 2 cycles per second at which the shell opens and closes in swimming.

Figure 4.16(b) shows an alternative arrangement, with springs in series with the muscles. As an example of this, the swimming muscles in the tails of dolphins have long tendons that must store and return energy as the tail beats up and down. In insects such as bees that have fibrillar flight muscles, the muscles themselves may serve as springs. Most muscles do not work effectively as springs because strain energy is stored mainly in the cross bridges and is lost whenever these detach and reattach, as the muscle lengthens and shortens. Bee muscle fibres, however, lengthen and shorten very little in flight, by only about 2 per cent of their length, and there may be no need for the cross bridges to detach. Unlike most muscles,

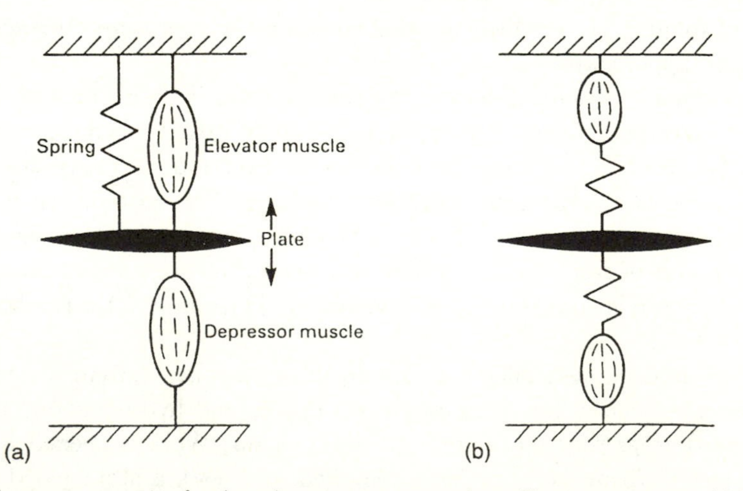

Fig. 4.16 Two arrangements of springs that can save energy in oscillatory movements: (a) has a spring in parallel with the muscles and (b) has springs in series with the muscles. (From R. McN. Alexander (1988) *Elastic Mechanisms in Animal Movement*, Cambridge University Press.)

the fibrillar flight muscles of bees and other insects may function as effective energy-saving springs as well as motors.

How can the energy cost of driving an oscillating system like the one shown in Fig. 4.16(b) be made as low as possible, by adjusting the properties of the muscles and springs? I have devised a simple computer model to try to find out. The model's muscles have properties as shown in Fig. 4.14, but the maximum shortening speed v_{max} can be varied. The compliance of the springs can also be varied (the compliance of a spring is extension divided by force, so a high compliance means that the spring is easy to stretch). The muscles are made to oscillate the plate sinusoidally with chosen frequency and amplitude. For every stage of the cycle of movement, the computer calculates the forces the muscles are exerting and the rates at which they are shortening or lengthening. Figure 4.14(a) is used to calculate the volume of muscle that must be active at each stage and Fig. 4.14(b) is used to calculate the metabolic rate per unit volume of the active muscle. From these data for all stages of the cycle, the total energy cost of the cycle is calculated. The results depend on the properties of the muscles and springs (their maximum shortening speeds and compliances) and on how large the inertial forces are relative to the hydrodynamic forces.

Figure 4.17 shows some results. In each graph, the vertical axis represents the maximum shortening speed of the muscles, divided by the maximum speed at which the plate moves in the course of a cycle. The horizontal axis represents the compliance of the springs; more precisely, it is the amount by which the peak hydrodynamic force would stretch the springs, divided by the amplitude of the plate's movements. The contours show the efficiency, defined as the work done against hydrodynamic forces divided by the metabolic energy used.

Ignore Fig. 4.17(a) for the present. In Fig. 4.17(b) the muscle has no need to exert inertial forces. A swimming scallop is effectively like this because the spring in its hinge joint balances the inertial forces. The contours show that for maximum efficiency (marked by a star) the muscles should be fairly fast and there should be no springs in series with them. Scallop muscles have no springs in series with them, and their v_{max} is close to what the theory predicts.

In Fig. 4.17(c) the peak inertial and hydrodynamic forces are equal. Their ratio (one) is higher than for swimming dolphins but lower than for hovering bees. The contours show that for maximum efficiency the muscles should again be fast, but that in this case they should have compliant tendons (or other springs) in series with them.

In Fig. 4.17(d) the inertial forces are very much larger than the hydrodynamic ones. Here we are approaching the situation of mammalian running in which large inertial forces act as the body is accelerated and decelerated, but hydrodynamic forces (air resistance) are relatively tiny. The optimum design has very slow muscles with tendon springs in series with them. The spacing of the contours shows that to minimize energy costs, it is important to get the compliance exactly right. If the compliance is right and hydrodynamic forces are very low, hardly any

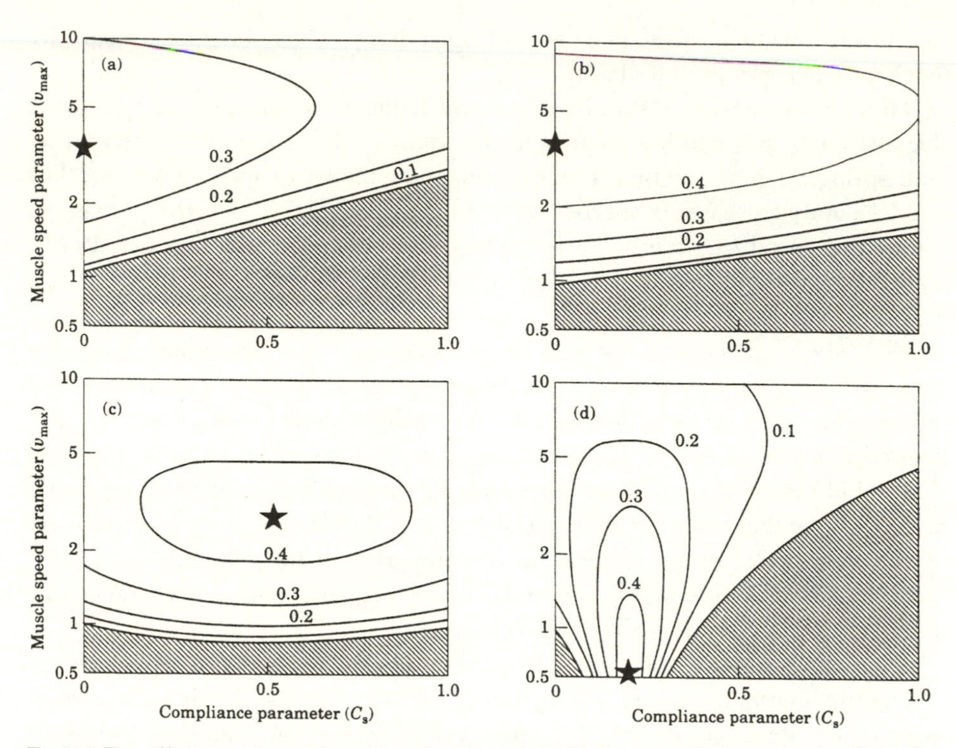

Fig. 4.17 The efficiency of muscles with springs in series with them oscillating a plate in a fluid (Fig. 4.16(b)). In each graph, contours show the efficiencies for different combinations of maximum shortening speed (v_{max}) and spring compliance. Graph (a) refers to a hypothetical situation, in which the inertial forces are effectively negative. In graph (b) there are no inertial forces, in (c) the peak inertial force equals the peak hydrodynamic force and in (d) it is five times the peak hydrodynamic force. Shading marks combinations of muscle speed and spring compliance that are impossible because the muscle would have to shorten faster than its v_{max}. (From R. McN. Alexander (1997) *Journal of Theoretical Biology* **184**, 253–9.)

work is required from the muscles to keep the system vibrating. Some of the main muscles in the lower parts of the legs of horses have long tendons that must serve as springs but muscle fibres only a few millimetres long that cannot shorten much or fast.

Figure 4.17(a) represents a situation in which the inertial forces are effectively negative. This would occur if the muscles were driving a resonant system at a frequency below its resonant frequency. This unrealistic situation is the only one of the four in Fig. 4.17 for which an efficiency of about 0.45 cannot be obtained.

4.5 **Sprinters and stayers**

We return now to fish. There is another difference between their two types of muscle, apart from their colour and speed. The red muscle works aerobically, obtaining its energy by oxidizing fats to carbon dioxide and water. It contains a lot of mitochondria, which house the enzymes of the Krebs cycle that plays the central role in aerobic metabolism. The white muscle works anaerobically, converting one molecule of glucose to two molecules of lactic acid (eqn 2.8). We have already seen this process happening in mussels exposed to the air at low tide (Section 2.7). It releases far less energy than aerobic metabolism would do; a molecule of glucose that would produce 38 molecules of ATP by aerobic metabolism provides only 2 when broken down to lactic acid. However, the rest of the energy is not lost, because subsequent oxidation of just some of the lactic acid can provide the energy to convert the rest back to glucose. The great advantage of anaerobic metabolism for fast locomotion is that the power output of the anaerobic muscles is not limited by the rate at which the respiratory system and the blood can supply them with oxygen. The fish can make a very rapid burst of activity, then take up oxygen and recover slowly.

Lactic acid production seems to have an important advantage over other anaerobic metabolic processes, such as those that produce succinate and propionate in tapeworms and mussels (Section 2.7); it seems to be capable of proceeding much faster.

Anaerobic activity cannot continue for long because the body tolerates only a limited concentration of lactic acid. That is why neither fish nor human athletes can sprint far at top speed. Fish can swim indefinitely in the range of speeds at which only the red muscles are used, but only for limited times at the higher speeds at which the white muscles are used. Men's world record running speeds are 10.2 m s^{-1} for 100 m, but only 7.6 m s^{-1} for 1000 m, and 6.1 m s^{-1} for 10 000 m. Similarly, a 20 cm trout was found to be able to swim at 1.8 m s^{-1} for 1 s but only at 1.2 m s^{-1} for 5 s and 0.6 m s^{-1} for 20 s.

The maximum running or swimming speed that can be sustained over long distances depends on how fast oxygen can be taken up and used by aerobic muscles. This depends in turn on how fast the respiratory system can take up oxygen, how fast the blood circulatory system can deliver the oxygen to the muscles, and how fast the aerobic muscles can use it. It would be pointless and wasteful, for example, to have a circulatory system capable of supplying oxygen much faster than the muscles could use it. The capacities of the respiratory, circulatory, and muscular systems should be well matched.

Professors Dick Taylor and Ewald Weibel led an ambitious research programme on the ability of mammals to take up and use oxygen when running. They confirmed that in general the systems involved are well matched. Figure 4.18 shows the metabolic rates of the animals when they were running on treadmills at the highest speeds they could sustain. If they ran faster they accumulated lactic acid and soon had to stop. Oxygen consumption was measured as shown in Fig. 4.1. The line shows that maximum metabolic rates tend to be proportional to (body mass)$^{0.81}$. In most cases they are about 10 times the minimal (resting) metabolic rate (see Fig. 2.3).

Figure 4.18 shows that domestic cattle and rats have lower maximum metabolic

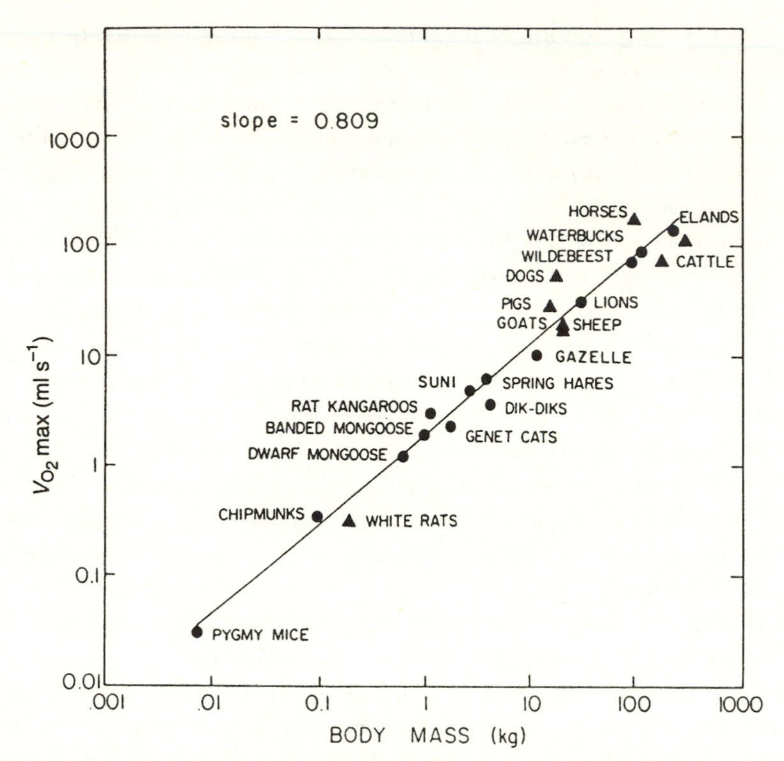

Fig. 4.18 Maximum rates of oxygen consumption during running of mammals ranging from mice to cattle. Consumption of 1 ml oxygen per second implies a metabolic rate of 20 watts. (From C. R. Taylor *et al.* (1980) *Respiration Physiology* **44**, 25–37. Copyright 1980, with permission from Elsevier Science.)

rates than would be expected for their masses, while dogs and horses have higher than expected maximum rates. These differences may have been enhanced by domestication; farmers breeding cattle have not cared whether they could run fast (they may even have preferred them to be slow), but people have been keen to breed fast horses. However, the most remarkable performance is that of a wild mammal, the pronghorn antelope (*Antilocapra*) of the North American prairies. Its maximum metabolic rate (not shown in the graph because it was measured in a later investigation) is 5 times that of a domestic goat of the same mass. This enables it to run many kilometres at 20 m s^{-1} (most horse races over one mile, 1609 m, or less are won at 16–17 m s^{-1}). There is no dramatic peculiarity to explain this achievement; the pronghorn has larger lungs than the goat, more haemoglobin in its blood, and more mitochondria in its muscles, but apparently nothing odder than that.

The proportions of aerobic and anaerobic fibres in muscles tend to be suited to the animals' ways of life. The white meat of a chicken's breast (the muscles that power flight) is anaerobic, and chickens make only occasional, very short flights. However, chickens spend a lot of their time running around, and the dark muscles of their legs are largely aerobic. Unlike chickens, pigeons make long flights and have dark, largely aerobic muscle on their breasts. Similar differences are found between human athletes from different sports. Both among runners and among cyclists, elite sprinters generally have high proportions of muscle fibres that can function anaerobically in their legs, while athletes that race over long distances have high proportions of aerobic fibres. Muscles respond to different training regimes in appropriate ways.

Few invertebrates seem to use anaerobic metabolism as vertebrates do, to enable them to sprint faster than they could do aerobically. However, the mantle muscles of squids (the muscles that power jet propulsion) consist of a small proportion of yellow aerobic muscle and a large proportion of white anaerobic muscle. The yellow muscle powers breathing and slow jetting, but the white muscle is recruited for bursts of fast swimming.

4.6 Journey planning

Travel uses energy and occupies time. Animals often have to decide whether a possible journey is worthwhile, or how best use can be made of a journey. These problems have been studied more for birds than for any other animals.

Consider a bird searching for food at a distance from its nest. It can carry only one food item at a time (perhaps an insect) back to the nest. Should it fly back to the nest with any insect it finds, or should it go on searching until it finds a big insect, ignoring the small ones?

Suppose for simplicity that the only choice is between taking every insect, or just the largest kind. Let the mean energy content of all the available insects be E_{av} and let the mean time required to find one be t_{av}. Large insects have energy content E_{big} (which is bigger than E_{av}) and take time t_{big} to find (which is longer than t_{av}). The two-way journey to the nest and back takes time T and uses energy PT (P is the extra metabolic rate during flight, additional to the rate when not flying). A bird taking every insect it finds gains energy $E_{av} - PT$ from each trip, which takes time $T + t_{av}$. One that takes only large insects gains energy $E_{big} - PT$ in time $T + t_{big}$. The one that takes only large insects will gain energy faster if

$$(E_{big} - PT)/(T + t_{big}) > (E_{av} - PT)/(T + t_{av})$$

which can be rearranged to give

$$T > (E_{av}t_{big} - E_{big}t_{av})/[E_{big} - E_{av} + P(t_{big} - t_{av})] \tag{4.2}$$

This tells us that if the bird is foraging far enough from its nest (T is large) it will be better to leave the small insects and take only the big ones, but that nearer the nest it may be better to take every insect it finds.

A version of this theory was tested in a study of bee-eaters (*Merops*) which had nests in burrows in the walls of ditches in Southern France. They were catching bees, dragonflies, and other insects and carrying them in their beaks back to young in the nest. One observer, in a hide at the top of a 4 metre tower, observed where each insect was caught. The other, hidden very close to the nest, identified the insect as best he could and estimated its size. Insects caught within 20 seconds' travel time of the nest were very predominantly small, and those caught beyond 60 seconds' travel time were nearly all medium or large. The travel time at which the birds stopped collecting small prey agreed quite well with the observers' initial theory, which took no account of the energy cost of travel. Puzzlingly, it agreed less well when they took account of energy costs.

We assumed in that discussion that the speed of the journey was fixed, so that the time needed to travel a given distance was predetermined. In fact, a bird has a choice of flying speeds. If the metabolic rate P is the same for all speeds, the bird should fly as fast as it can to cut down the journey time. If the metabolic rate/speed curve is U-shaped (Fig. 4.8(b)) the bird can minimize the energy PT used on each journey by flying at its maximum range speed. However, it will be able to increase its rate of energy gain by flying a little faster, using more energy on each trip but fitting more trips into the day.

Some of the best information about bird flight speeds comes from radar tracking of seabirds. The velocities measured from a ship were corrected for the velocities of the wind and of the ship to obtain speeds relative to the air. Wind can affect optimum flight speeds, so I will give data only for near-calm conditions. Minimum power and maximum range speeds were estimated for each species from their wing dimensions, but the method of calculation depended on an assumption about the aerodynamic drag on the body which has since been shown to be false. The minimum power and maximum range speeds given below are corrected values. Most of the species did a lot of gliding, but four flapped their wings continuously. For one of these flapping species (the little shearwater, *Puffinus assimilis*) the average speed of 14 m s^{-1} was intermediate between the minimum power speed of 11 m s^{-1} and the maximum range speed of 18 m s^{-1}. Imperial cormorants (*Phalacrocorax atriceps*) and south polar skuas (*Catharacta maccormicki*) flew at mean speeds slightly below their minimum power speeds of 18 and 14 m s^{-1}, respectively. Wilson's petrel (*Oceanites oceanicus*) flew about at 6 m s^{-1} when it was hunting for food on the sea surface, and at about 9 m s^{-1} when it was travelling between patches of food, just below and above its minimum power speed of 8 m s^{-1}. These and other data show that birds commonly fly at speeds close to their minimum power speeds, despite the theoretical prediction that they could travel the same distances at less energy cost if they flew faster, at the maximum range speed. In some cases the reason may be that they cannot fly faster, because the maximum aerobic power output of their flight muscles may be too low for sustained flight at the maximum range speed.

Many birds and some other animals make long annual migrations, spending different seasons at places thousands of kilometres apart. For example, many birds that spend the summer in Europe spend the European winter in Africa south of the Sahara. When is a migration worthwhile?

Suppose that a bird spends the (northern) summer in the northern hemisphere. If it cannot find enough food to survive in the northern winter it must migrate, and an insect-eating bird might find it especially difficult to survive because insects are so much less plentiful in winter. We will suppose, however, that survival over the winter is possible, without migrating. Let τ be the duration of the winter. During this time, the bird can take in food energy at a rate F_N if it remains in the northern hemisphere or at a greater rate F_S if it moves to the southern hemisphere. The round trip takes time T, during which the bird uses energy at a rate P in excess of what it would use if it were not travelling. If the bird remains in the northern hemisphere all winter it will gain energy $F_N\tau$. If it travels to the southern hemisphere it will lose energy PT on the journey and its feeding time will be reduced to $(\tau - T)$. Its energy gain will be $F_S(\tau - T) - PT$. The migration will be worthwhile if

$$F_S(\tau - T) - PT > F_N\tau$$

$$F_S > (F_N\tau + PT)/(\tau - T) \tag{4.3}$$

This shows that the time and energy needed for the journey can affect the profitability of a migration, but let us ask how important they are likely to be. Suppose that the round trip distance is 10 000 kilometres, which is about the distance for many migrants between Europe and Africa. Many birds fly at around 10 metres per second, at which speed they would spend 24 days each year travelling if they flew 12 hours per day. They could travel faster with a following wind, but I am assuming that any advantage from the wind on the outward journey is cancelled out by headwinds on the return journey. Twenty-four days is about 0.13 of a six-month winter, so we assume $T/\tau = 0.13$. While flying, birds use energy at about 10 times the minimal rate. During 24 h periods in which they fly all day and rest all night the mean metabolic rate will be about five times the minimum value. Birds that are not migrating would probably metabolize at 2.5–3 times the minimum rate, as will be shown for house martins in Chapter 7. Thus P is 2–2.5 times the minimum rate and we will suppose that it is 1.5 times F_N. With these assumptions, the theory predicts that migration would be worthwhile if F_S was at least 1.4 times F_N. Food would have to be at least 1.4 times as plentiful in the southern hemisphere habitat as in the northern one, during the northern winter. For many birds, it seems likely that it is far more than 1.4 times as plentiful. There seems little doubt that these long migrations are highly profitable for many bird species.

Now suppose that a terrestrial mammal were considering making a 10 000 km round trip migration. Assume (improbably) that there are no seas, deserts, or mountain ranges, which would be formidable barriers to a terrestrial mammal but might give little trouble to a flying bird. It would travel far more slowly than a bird, perhaps at 3 metres per second, which is a jogging speed for a human, a trotting speed for a pony, and a galloping speed for a rat. At that speed, again travelling for 12 hours each day, the travelling time would be 80 days. It seems most unlikely that it would be worthwhile, to spend so much of the year travelling between feeding areas.

No terrestrial animals make really long migrations; the longest round trips are about 1000 km for wildebeest (*Connochaetes*) in the Serengeti and 2000 km for caribou (*Rangifer*) in Canada. However, some swimming animals make remarkable migrations. For example, humpback whales (*Megaptera*) migrate annually around 7000 km each way between the Antarctic and the tropics.

Another question that arises in locomotion is, which is the best route? For many of the storks migrating between Africa and Europe the most direct route would be over the wide part of the Mediterranean, but most of the birds go round by Suez or Gibraltar, so that their journey is almost entirely over land. This saves energy by enabling them to soar all the way—there are no thermals over the Mediterranean.

This chapter has reviewed the ways in which animals run, fly, and swim, looking particularly at energy costs. It has shown that some techniques of locomotion are more economical than others; for example, fishes swim more cheaply than squids. Also, many animals save energy by changing gaits as they increase speed. We have seen how the shortening speeds of muscles and the elastic properties of their tendons can be adapted to minimize energy costs. It has been shown that anaerobic metabolism enables some animals to sprint much faster than would be possible if they were limited by the rate at which they can supply their muscles with oxygen. And we have examined some questions of energy raised by journeys, both short foraging trips and long migrations.

5 Growth and reproduction

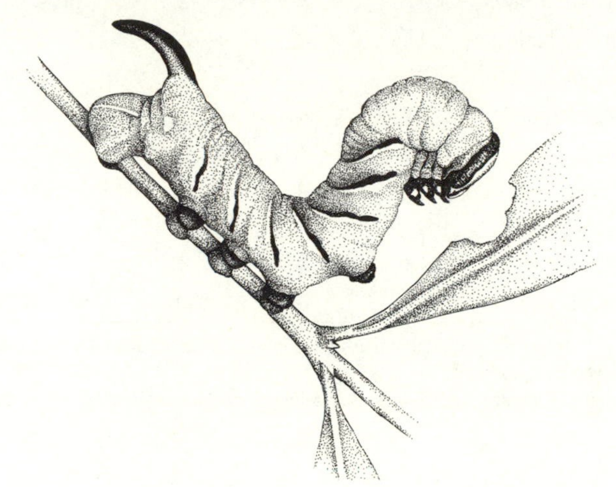

This chapter examines the energy costs of growth and reproduction. It asks, what can be achieved, with a given amount of energy? When is it better to use available resources for growth, and when for reproduction? If for reproduction, how can the resources best be used—to produce a few large eggs or many small ones, to produce males or females, and so on?

5.1 Energy balance

As an animal grows, it adds to the energy content of its body. Any energy gained from food and not used in any other way can be retained for growth:

$$\text{food energy} - \text{excreted energy} - \text{heat stored or lost} - \text{work done} = \text{energy retained} \tag{5.1}$$

We will see that this is a modified version of eqn 2.2. The principle of conservation of energy tells us that it must balance. By food energy I mean the enthalpy of combustion of the food, H_{food} in eqn 2.2. Excreted energy is the enthalpy of combustion not only of the faeces but also of the ammonia, urea, and other compounds in the urine and, if the animal ferments its food, any methane or other combustible gases that are produced. In eqn 2.2 it is represented by $H_{\text{faeces}} + H_{\text{urine}}$. The term **metabolizable energy** is often used to mean food energy minus excreted energy. Some examples, given in Table 5.1, show that metabolizable energy may range at least from 30 per cent to 90 per cent of food

Table 5.1 Energy losses in faeces, urine, and methane and the metabolizable energy of various diets, expressed as percentages of their enthalpies of combustion

Species	Diet	Faeces	Urine	Methane*	Metabolizable energy
Horse	Hay and oats	35	4	2	59
	Wheat straw	61	4	3	32
Sheep	Barley grain	17	3	11	70
	Barley straw	61	2	7	31
Pig	Maize grain	14	4	?	82
	Kale	34	4	0.4	62
Human	Low fibre	4	4	?	92
	High fibre	8	3	?	89
Rat	Chow	13	4	0	83
Hen	Layer ration	17	3	0	81

*In cases in which methane production was not measured it was probably very small, so little error has been introduced by ignoring it.
Based on a table in K. Blaxter (1989) *Energy Metabolism of Animals and Man*, Cambridge University Press.

energy. The range is wide mainly because some diets are broken down in the gut much more completely than others, as discussed in Chapter 3.

The heat in the equation is either retained in the body, increasing its temperature, or lost to the environment. It may come from metabolism, or may be released from food in the gut in the course of digestion or fermentation. Heat loss here means net loss, after taking account of any heat gained from the environment, for example when the sun's rays warm the animal.

The work in the equation might be the work done on air or water in the case of flight or swimming, the hydrodynamic work of Section 4.4. Alternatively, it might be the work done against gravity while running up a mountain. The work done giving the body potential and kinetic energy while running over level ground does not appear here because potential and kinetic energy supplied by muscular work at one stage of a stride is removed later in the stride by muscles acting as brakes, which convert it to heat. For a similar reason, inertial work done in flying or swimming does not appear. In any case, work is never more than a small fraction of the heat produced by metabolism because muscles generally work with efficiencies of 25 per cent or less: for every joule of work that muscles do at least three joules are lost as heat. The heat plus the work in eqn 5.1 is represented by M in eqn 2.2.

Energy retained may be the enthalpy of combustion of material added to the body by growth or in laying down food stores such as fat deposits. It may also be energy retained for reproductive purposes, in developing ovaries or testes, in a fetus developing in the womb, or in milk produced by a lactating mother. It is represented by H_{growth} in eqn 2.2.

In this chapter we focus on the retained energy. What are the energy costs of growth and reproduction? If only limited energy is available, when is it best to use it for growth and when for reproduction?

5.2 Costs of growth

The energy cost of growth is not merely the enthalpy of combustion of the proteins and other materials that are added to the body; additional energy is needed, which gets lost as heat. Look again at Fig. 3.4(c), which shows the growth rates of mussels growing in algal suspensions of different concentrations. Energy incorporated in the body by growth is plotted against the rate of intake of metabolizable energy, which is low in dilute algal suspensions and high in concentrated ones. When food intake is very slow, growth is negative; some of the body's energy is used in metabolism and not replaced. A rate of intake of 3 joules per day (metabolizable energy) was just enough to balance metabolism and the energy content of the mussels remained constant; they neither grew nor shrank. Higher rates of intake enabled the mussels to grow; the faster the intake the faster the growth. Any spare energy, over and above maintenance requirements, was used for growth.

The gradient of the steepest part of the line in Fig. 3.4(c) is 0.7, telling us that the energy incorporated by growth is 70 per cent of the spare metabolizable energy. Putting the same thing differently, the energy requirement for growth is $(1/0.7) = 1.4$ times the energy incorporated. We saw in Section 3.5 that the spare energy that was not incorporated was used partly for the processes of digestion, but probably mainly for protein synthesis.

Figure 5.1 shows the results of an experiment on cattle, similar to the experiment on mussels. Each animal was given the same, precise amount of food daily

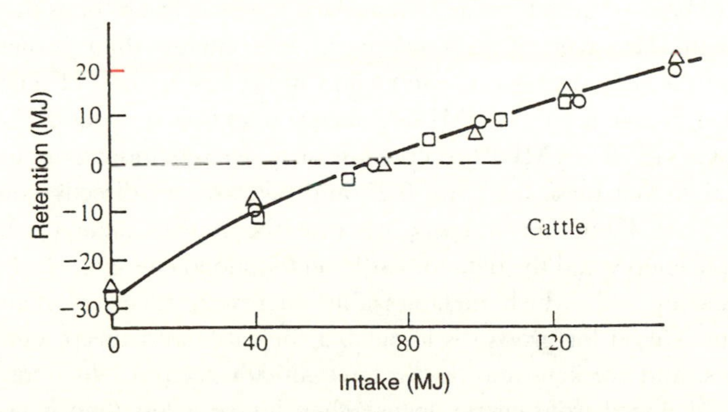

Fig. 5.1 Energy retention plotted against the rate of food energy intake, for cattle given different daily rations of food. (From K. Blaxter (1989) *Energy Metabolism in Animals and Man*, Cambridge University Press.)

for 3 weeks, and measurements were made over the last 4 days. The long wait before starting the measurements was needed because food takes a very long time to pass through the guts of cattle; the mean retention time is 60 hours, and some of the food takes much longer. If the measurements had been taken while some of the food from a different diet was still being processed, the results would have been misleading.

Energy incorporated by growth in experiments like this cannot be measured simply by weighing the animal before and afterwards. An animal that adds a kilogram of fat to its body adds much more energy than one that adds a kilogram of protein, and one that drinks a kilogram of water adds no energy at all. In the experiments on mussels, animals were dried and then burned in a calorimeter to measure their enthalpy of combustion, but it would be a daunting experiment to burn a complete cow and measure the heat produced. An alternative is to dissect the animal after the experiment into components such as muscle and fat; weigh each component; and measure the enthalpy of combustion of a small sample of each. The errors resulting from this sampling procedure will generally be less than the errors resulting from the problem that you cannot determine the animals' enthalpies of combustion before the period of growth. The initial enthalpy of combustion has to be estimated from measurements of similar animals killed at the beginning of the experiment. There are other ingenious methods for estimating the energy content of the body which are especially useful for experiments on humans, who cannot be slaughtered even at the end of an experiment.

Note that the rates of energy intake in Fig. 5.1 are total intake of food energy, not just metabolizable energy as in the experiments on mussels. The cattle kept their energy content constant when they were fed at the rate of 70 megajoules per day (MJ d^{-1}). With more food the energy content of the body rose and with less it fell. The graph is quite evidently curved. The reason seems to be that if an animal is taking in more energy than is needed for maintenance, some of the surplus energy has to be used both for digestion and for the protein synthesis that accompanies growth. However, if it is taking in less energy than is needed for maintenance the only saving is the reduction in the energy cost of digestion. In Fig. 5.1, when intake is $(70 + x)$ MJ d^{-1}, energy retention is about $0.25x$ MJ d^{-1}; but when intake is $(70 - x)$ MJ d^{-1}, energy loss (negative retention) is about $0.4x$ MJ d^{-1}. Remember that these fractions, 0.25 and 0.4, are not directly comparable with the 0.7 we found for mussels, because the mussel data are based on metabolizable energy and the data for cattle on total food energy.

The efficiency with which metabolizable energy in excess of maintenance requirements is used for growth is about 0.5 for cattle and sheep, but 0.75 for humans, rats, and chickens and (as we have already seen) 0.7 for mussels. The 'efficiency' calculated from energy losses when intake is less than is needed for maintenance is about 0.7 for cattle and sheep and 0.9 for humans, rats, and chickens.

5.3 Costs of reproduction

Just as the total energy cost of growth is greater than the enthalpy of combustion of the new tissues, the energy cost of producing eggs, fetuses, or (presumably) sperm is greater than the enthalpy of combustion of these products.

The energy cost of egg production has been studied most intensively for domestic hens. In the laying season, a 2 kg hen typically produces one 60 gram egg each day. This is 3 per cent of body mass, but a much smaller proportion of the energy content of the hen because the enthalpy of combustion of the egg (including the low-energy-content white as well as the high-energy-content yolk) is only 6.2 MJ kg^{-1}, compared to 16.7 MJ kg^{-1} for the bird. The enthalpy of combustion of a 60 gram egg is about 370 kJ and the minimal metabolic rate of a 2 kg hen is about 562 kJ d^{-1}, so the energy stored in eggs, when one is produced each day, amounts to two thirds of the minimal metabolic rate.

The cost of storing this energy has been estimated by measuring the metabolic rates of laying and non-laying hens. It appears that a hen that is getting the energy for egg production from its food, rather than by drawing on the energy reserves in its body, converts metabolizable energy of food to enthalpy of combustion of eggs with an efficiency of about 0.65. Thus the cost in metabolizable energy of producing a 370 kJ egg is 570 kJ, about equal to the minimal daily metabolic rate.

The efficiency of egg production is quite similar to the efficiency of growth (about 0.75 for chickens, Section 5.2), which should not surprise us. However, if we calculate the efficiency of producing a hatchling instead of just an egg, the efficiency is much lower because some of the energy of the egg is used for the metabolism of the developing embryo and because the eggshell and the allantochorion fail to get incorporated into the embryo. (The allantochorion is a membrane with a rich blood supply which develops close under the shell and serves the vital function of acquiring the oxygen needed for the embryo's metabolism.) Of the 370 kJ enthalpy of combustion of a typical newly laid egg, 100 kJ is used for metabolism during incubation, 130 kJ is eventually discarded in the shell and allantochorion, and only 140 kJ remains in the newly hatched chick.

A pregnant mammal deposits energy in the placenta as it develops (mainly in the early stages of pregnancy) and in the fetus. For example, in the case of a 70 kg sheep producing triplets of total mass 12 kg, the enthalpies of combustion at full term are 12 MJ in the placenta and so on (including the uterus and fetal fluids) and 52 MJ in the fetus. This total of 64 MJ is laid down during a pregnancy of almost 150 days, but the rate of storing energy increases rapidly towards the end of pregnancy and is 2 MJ d^{-1} in the final stages. This is 40 per cent of the minimal metabolic rate.

A mammalian mother continues to supply energy to her young after birth, as milk. The efficiency of milk production has been measured by experiments in which cattle have been fed different rations while growth and milk production were measured. When food intake is fast enough to supply the energy for milk production as well as for maintenance, dairy cattle convert metabolizable energy to milk with an efficiency of about 0.6; but when food is short and the energy for milk production comes from the body's reserves the efficiency is about 0.8. The efficiency of 0.6 is rather better than the corresponding efficiency of 0.5 for growth (see Section 5.2).

A calf may use the energy of milk either for metabolism or for growth. Milk is highly digestible so nearly all its energy is available to the calf; 95 per cent of its energy content is metabolizable energy. However, because cows produce milk with only 60 per cent efficiency, the metabolizable energy that the calf gets is much less than the cow could have got from the same quantity of grass.

Further energy loss occurs when the calf uses the milk as the energy source for growth. This has been investigated by studying the growth of calves given different daily rations of milk. A calf needs a certain amount of milk simply to maintain its body mass. It has been found that once this has been supplied, any additional energy is incorporated into the body by growth, with an efficiency of about 0.8. Thus the efficiency of conversion of surplus metabolizable energy in the cow's diet to growth of the calf is about $0.6 \times 0.8 \approx 0.5$. Energy in the cow's diet is used for calf growth as efficiently as the cow could use it for her own growth. These comparisons make milk seem rather an inefficient means of supplying energy for metabolism, but a remarkably efficient way of supplying it for growth.

There are other energy costs for parents. Chapter 7 gives information about the energy used by worker bees collecting food for the young in the nest and by parent birds collecting food for their nestlings.

5.4 **Investment in growth or reproduction**

Growth and reproduction both require energy and materials. The quantity of energy available to an animal is always limited, so how should animals use any surplus that is not needed for maintenance? When is it better to use the available energy for growth, and when for reproduction?

Evolution by natural selection favours those genotypes (sets of genes) that have the highest fitness; that is, that have the highest probability of transmission to subsequent generations. If a particular set of genes is possessed by n_1 members of one generation and n_2 members of the next (counted at the same stage in the life history), its fitness is n_2/n_1. Thus natural selection will favour genotypes that produce many offspring over those that produce few. It is not merely a matter of the number of eggs or newborn offspring—a genotype that produces a small number of large offspring that survive well may have higher fitness than one that produces a large number of small offspring that are less likely to survive, but that is the subject of another section. In this section we will assume that the size and probability of survival are fixed and the problem is simply to produce as many offspring as possible.

Obviously, an animal can maximize its current rate of reproduction by putting all available energy into reproduction and none into growth. However, by investing energy in growth now it will become bigger, which may enable it to obtain more energy in the future; and by delaying reproduction until then, it may eventually be able to produce more offspring. For example, a large fish can lay many more eggs in a season than a small one could; in general, a female fish that is twice as heavy as another of the same species will lay twice as many eggs.

The point that it may be best to grow first and reproduce only later can be demonstrated mathematically, but the proof is rather complicated. It seems easier to make the point here by using a very simple example. The example may seem like cheating because it is not about an individual animal's choice between growth and reproduction, but about the choice faced by a colony of wasps.

Some wasp species are solitary, but we are concerned here only with the ones that form colonies. Each colony includes members of several different castes. Commonest are the workers, females that do not reproduce but undertake the work of collecting food and caring for the eggs and young. There are also reproductive males and females. In temperate climates, the only wasps to survive the winter are the reproductive females that will found new colonies in the spring and become queens. They will have copulated in the previous autumn with males, whose sperm they store and use throughout the season. In the spring, when she wakes from her winter torpor, the queen starts her colony by laying a few eggs which hatch and develop to become workers. Once these are adults, there is strict division of labour; the queen lays eggs and the workers do all the rest of the work.

Later in the season, the queen lays some of her eggs in larger cells and these eggs develop to become reproductives. Whether they are male or female is not a matter of chance, as in humans and most animal species, but depends on whether the queen fertilizes them or not; we will discuss that complication later. The reproductives do no work.

Most of the reproductives will perish without parenting new colonies, but some will survive to become queens or to fertilize queens. Other things being equal, the more reproductives a colony produces the better its chance of passing its genes to next year's colonies. To maximize its fitness, the colony should produce as many reproductives as possible. That might suggest that the colony's best policy was to devote all the available energy to producing reproductives; but in that case the queen would have no workers to feed and care for the young. She would have to do all the work herself and could manage to rear only a few offspring. She can produce more reproductives in the end if she first builds up a large team of workers.

We will work out the best strategy, making very simple assumptions that are based on observations of a hornet (a large wasp, *Vespa orientalis*) in Israel. Colonies of this species have approximately 10 members on 15 May, and continue reproducing until cool weather sets in in mid November. The rate of reproduction is limited by the rate at which the workers can collect food for the young, not by the rate at which the female can lay eggs; she lays as many eggs as are needed, one in each larval cell that the workers build. Each worker can provide enough food to rear one new individual each month, either a worker or a reproductive. Thus if there are *n* workers at the beginning of the month, *n* new individuals can be produced in the course of the month, and these may be all workers, or all reproductives, or a mixture of the two.

There would be no point in producing reproductives first and only workers later, because the workers would then do nothing to increase the output of reproductives, on which the fitness of the colony depends. We will assume that the colony produces workers first and reproductives later, switching abruptly from one to the other. Table 5.2 shows the effects of switching on different dates.

As already specified, there are 10 workers on 15 May. If only workers are produced, there will be 20 on 15 June, 40 on 15 July, and so on. If the number can be doubled in a complete month it can be multiplied by 1.4 (the square root of two) in half a month, so there would be 14 on 1 June, 28 on 1 July, and so on.

The third column shows the number of months remaining until reproduction ceases at the onset of winter, in mid November. The final column shows for each date the number of reproductives that would be produced by the end of the season, if the switch from producing workers to producing reproductives were made on that date. The numbers in this column have been obtained by multiplying the numbers in the previous two columns. For example, if production of reproductives starts on 15 September, the 160 workers present on that date can produce 160 reproductives in each of the 2 months that remain, a total of 320 reproductives. However, the largest number of reproductives is obtained by

Table 5.2 Calculation of the numbers of offspring produced by a hornet colony that switches from producing workers to producing reproductives on each of the dates shown (further explanation is given in the text)

Date	Final number of workers	Months remaining	Final number of reproductives
15 May	10	6.0	60
1 June	14	5.5	77
15 June	20	5.0	100
1 July	28	4.5	126
15 July	40	4.0	160
1 August	56	3.5	196
15 August	80	3.0	240
1 September	112	2.5	280
15 September	160	2.0	320
1 October	224	1.5	336
15 October	320	1.0	320
1 November	448	0.5	224

switching half a month later, on 1 October. The Israeli hornets on which the calculation is based did in fact switch from producing workers to producing reproductives on or about 1 October. It must be partly good luck that this very simple theory predicts the switching date so accurately—a version of the theory that is more realistic, in that it takes account of the probability that some of the workers will die before the end of the season, actually predicts the date less accurately.

So far we have ignored the possibility that it might be possible to do even better by producing workers and reproductives simultaneously for at least part of the season. Suppose for example that half the workers present on 1 September switched then and half delayed switching until 1 October. The final number of reproductives would be the mean of the numbers that would be produced if all the workers switched on 1 September and if all switched on 1 October, which is 308. Whatever pair of switching dates we chose, the mean for the two dates would always be less than the 336 reproductives produced if all the workers switched their efforts on 1 October. The best strategy is a sudden, complete switch—what control theorists call a 'bang–bang' strategy.

A wasp colony is in some ways rather like an individual animal of a non-colonial species. Adding workers to a colony is equivalent to growth of an individual animal, and adding reproductives is investing in reproduction. We can apply almost exactly the reasoning we used for the colony to an individual animal, and ask when it should use the energy it has for growth, and when for reproduction. To keep things really simple, assume that we are considering a species that has only one generation each year, with all the adults dying at the end of the season. Assume also that big animals can collect food faster than small ones, making the energy available each month for reproduction or growth proportional to body mass at the beginning of the month. Then an argument like the one for wasp colonies will show that to maximize fitness, the animal should grow without reproducing in the early part of the season, then reproduce without growing in the later part. There are both animals and plants that behave like this. Insects grow as they pass through successive larval stages then, as adults, reproduce without further growth. Annual flowering plants grow in the early part of the season, then flower and set seed in the later part.

That argument ignores the possibility that by reproducing earlier the animal may achieve a shorter generation time. To explore this possibility, make a different set of assumptions. Suppose that breeding always happens at one particular season, but that this may be either when the animal is 1 year old or when it is 2 years old and that death immediately follows reproduction. First consider a genotype that breeds at 1 year old. An animal born in year 0 has a probability s_1 of surviving to the breeding season in year 1, when it will be able to produce n_1 offspring. Allowing for the possibility that it will not survive, the expected number of offspring is $s_1 n_1$. If the survivors of these breed in year 2,the expected numberof offspring in year 2 is $(s_1 n_1)^2$.

Now consider a genotype of the same species that breeds at 2 years old. If it is born in year 0 it has a probability s_1 of surviving to year 1. The probability of surviving from then to year 2 is s_2 and (if it survives) it will be able to produce n_2 offspring. Its expected number of offspring (taking account of the possibility of death in both years) is $s_1 s_2 n_2$. It is better to delay breeding to year 2 if

$$s_1 s_2 n_2 > (s_1 n_1)^2$$

$$s_2 / s_1 > n_1^2 / n_2 \qquad (5.2)$$

In that argument I have assumed that offspring are identical, whether produced by a 1 or 2 year old parent. Animals in their first year of growth will be smaller and more vulnerable than in the second so s_2 can be expected to be larger than s_1. Two-year-old parents who have devoted an extra year's spare energy to growth will be bigger than 1 year old parents and will be able to produce more offspring, so n_2 will be greater than n_1, but it may not be larger than n_1^2. Thus condition 5.2 may or may not be satisfied. In some cases the annual strategy will be best but in others the biennial one may be better. In yet other cases it may be best to grow for several years before breeding. Among insects, the goat moth (*Cossus*) spends three or more years as a caterpillar before pupating and becoming adult.

We have been assuming that the animals breed only once and then die, but many animals breed repeatedly. Mathematical analysis of that kind of life history is unfortunately a good deal more difficult and I will avoid it here (but I have explained how it can be done in my book *Optima for Animals* listed in the Bibliography).

Many animals grow to adult size, then cease growth and start breeding; mammals and birds are familiar examples. Our discussion has shown (for very simple cases) why that may be a good strategy. However, many fishes continue to grow over several seasons of breeding. Nothing discussed so far explains why that might be good, but a possible explanation emerges when we consider the effect of reproduction on survival. Again, mathematical analysis is difficult (and again, *Optima for Animals* explains how it can be done). It depends on the assumption that the faster an animal produces offspring, the more likely it is to die. This has been shown to be true for various animals. One example is the rotifer *Asplanchna*, which breeds parthenogenetically in suitable conditions, with no need for males. In such conditions, some females breed much faster than others. It has been shown by keeping females in separate dishes that fast breeders, producing on average 4.2 offspring daily, lived on average for 2.5 days; but slow breeders, producing 3.2 offspring daily in the same conditions, lived for 4.0 days.

The mathematical analysis shows that it is bad policy to put all available energy into breeding, at any stage of life, if a graph of mortality against rate of reproduction curves upwards; that is, if doubling the rate of reproduction more than doubles the death rate due to reproduction. In such a case, it is better to devote only some of the available energy to reproduction and the rest to growth.

That is what many fish do, but Pacific salmon (*Onchorhynchus*) breed once and then die. They spend several years of their life in the sea, then migrate up a river to breed. This is a long and strenuous journey, against the current. It has to be made, whether the fish produces a lot of offspring or only a few. A female that laid only a few eggs would be almost as exhausted, and almost as likely to die, as one that laid the maximum possible number. Doubling the number of offspring will presumably less than double the mortality, and the best strategy for reproduction is a single, all-out effort.

5.5 Use of resources for reproduction

In the previous section we asked what proportion of the available energy an animal should invest in reproduction at each stage of its life, and what proportion in growth? Now we ask, once the allocation to reproduction has been decided, how can it best be used? Is it better to produce a large number of small eggs or a few large ones? Is it better to have equal numbers of male and female offspring or to produce predominantly one sex?

First, the question of small eggs or large. If the energy available for producing eggs is E, and n eggs are produced, the energy content of each individual egg is E/n. The bigger the egg, the larger and less vulnerable the newly hatched larva will be and the less time it will need to grow to adult size; a larva from a big egg is more likely to survive to adulthood. The options are to produce a large number of small eggs, each with a low probability of survival; or a few large eggs with a higher probability of survival.

Figure 5.2(a) is a schematic graph of probability of survival P against the energy content of each egg, E/n: think of this energy content as a measure of egg size. The graph can be expected to curve, as shown, for the following reasons. First, there must be a minimum egg size below which survival is impossible. Second, it seems likely that once a certain size has been reached, further increase will make rather little difference to the probability of survival. We can use graphs like this to predict an optimum egg size.

Any point on the line in Fig. 5.2(a) has coordinates (E/n, P). The gradient of a

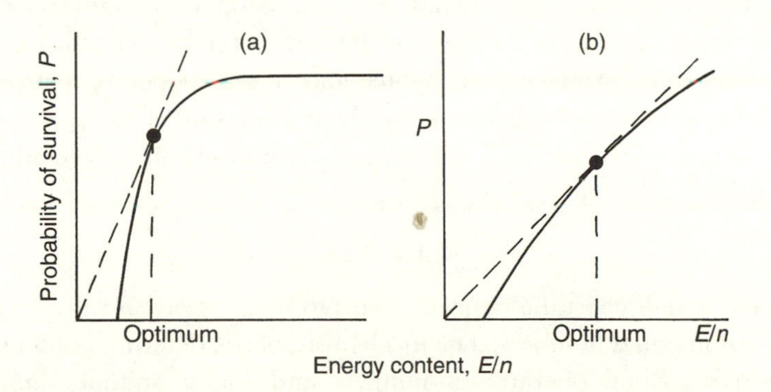

Fig. 5.2 (a), (b) Schematic graphs of probability of survival to maturity against egg size.

line from the origin, through this point, is Pn/E. The tangent shown in the figure is the steepest such line, and its gradient is the largest attainable value of Pn/E. Natural selection will tend to maximize the number of eggs expected to survive. This is Pn, the number of eggs multiplied by the probability that each will survive. The energy E is assumed constant. Thus the egg size at which the tangent touches the curve, which gives Pn/E its greatest value, is the optimum.

Figure 5.2(b) resembles 5.2(a) except that the graph is less strongly curved: in case (a), above a certain size, large eggs are only a little more likely to survive than small ones; but in case (b), survival continues to increase quite steeply as size increases, even at very large sizes. The tangent in Fig. 5.2(b) shows that in this case, optimum egg size is larger than in Fig. 5.2(a).

The conclusion from this is that if large eggs are a lot more likely to survive, it may be best to produce a few large eggs, but if large eggs are only a little more likely to survive it may be better to produce a lot of small ones. This seems to be demonstrated by guppies (*Poecilia*) living in streams in Trinidad. These fish do not lay eggs but give birth to larvae, which does not change the principle. In one stream, the principal predator was *Crenicichla*, a fish that ate mainly adult guppies. In another, the predator (again a fish) was *Rivulus*, which ate juvenile guppies. The population in the *Rivulus* stream (where small larvae were in more danger) produced fewer, larger larvae than the one in the *Crenicichla* stream. This is what the theory would lead us to expect. However, in applying the theory we should remember that evolution seems often to be constrained by its starting point. Selachians such as dogfish and rays lay much larger eggs than teleosts do, in the egg cases known as mermaids' purses. It seems possible that a dogfish might have higher fitness if it laid thousands of tiny eggs, like many teleosts; but if intermediate-sized eggs are less favourable than very large or very small ones, it will not be possible for a large-egged selachian to evolve tiny eggs.

Now suppose that an animal has the choice of using its limited resources to produce male or female offspring. Let the energy cost of producing a male be C_m, and the cost of producing a female be C_f. Let a population of n_m males and n_f females produce N offspring, by sexual reproduction. Then males will have on average N/n_m offspring and females N/n_f. An animal that invests energy C_m producing a son can expect N/n_m grandchildren, a return per unit investment of N/n_mC_m. One that produces a daughter gets a return per unit investment of N/n_fC_f. Natural selection will favour production of sons if $n_mC_m < n_fC_f$, and of daughters if the reverse is true. Consequently, evolution will tend to adjust the sex ratio to make

$$n_mC_m = n_fC_f \tag{5.3}$$

If males and females cost the same, as seems to be the general rule, they will then be produced in equal numbers. The mechanism of sex determination by X and Y chromosomes, which operates in humans and many animals, automatically produces equal numbers of the two sexes.

The question of the optimum sex ratio becomes more interesting when the Hymenoptera (ants, bees, and wasps) are considered. In this group, fertilized eggs become females and unfertilized ones become males. Also, in social species the workers may be able to decide the sex ratio by caring preferentially for one sex. These peculiarities have a profound effect on sex ratios.

Females, having developed from fertilized eggs, are diploid, with a full double set of chromosomes. Males, from unfertilized eggs, are haploid, with only a single set of chromosomes. Generally, all the young produced in a colony have the same mother (the queen). Females (both workers and reproductive females) all have the same father. However, male reproductives have no father. Thus a worker and a reproductive female reared by her will share on average half of the genes that the worker inherited from their diploid mother and all the genes that she inherited from their haploid father: in all, three quarters of the worker's genes. In contrast, a worker and a male reared by her share only one quarter of the worker's genes: half of the genes that she inherited from their mother. The reproductive female has three times the potential of the male for carrying the worker's genes to the next generation. Modifying eqn 5.3 accordingly, we conclude that workers producing reproductive females or producing males will get equal value (in transmission of genes matching their own to the next generation) from their investment, if

$$3n_{\mathrm{m}}C_{\mathrm{m}} = n_{\mathrm{f}}C_{\mathrm{f}} \tag{5.4}$$

Natural selection can be expected to adjust the sex ratio to make it satisfy this equation.

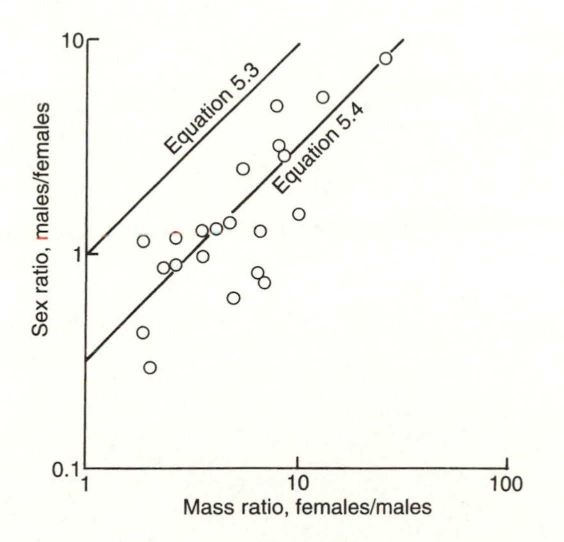

Fig. 5.3 Sex ratios in colonies of 21 species of ant. Sex ratio ($n_{\mathrm{m}}/n_{\mathrm{f}}$) is plotted against the ratio of adult body masses (which is assumed to be the ratio of costs $C_{\mathrm{f}}/C_{\mathrm{m}}$). The lines represent eqns 5.3 and 5.4. (Data from R. L. Trivers and H. Hare (1976) *Science* **191**, 249–63.)

This theory has been tested by digging up complete colonies of ants and counting and weighing the reproductives. The females were larger than the males in every case, ranging in different species from about two to 25 times the males' body mass. They grew to these masses on food provided by the workers, which implies that the ratio of costs to the workers, C_f/C_m, ranged from two to 25. In some species there were many more males than females, and in others the reverse was true. Figure 5.3 shows that the data fit eqn 5.4 reasonably well.

In this chapter we have focused on the energy available to animals for growth and reproduction. We examined the efficiency of growth, finding that surplus food (that is, in excess of maintenance requirements) cannot all be incorporated in the body; some has to be used to supply energy for protein synthesis. We went on to consider the efficiencies of egg production, pregnancy, and lactation. Next we asked how limited resources are best divided between growth and reproduction. Often, the best strategy is to put all available resources into growth at first, then to stop growing and reproduce. However, in some circumstances it can be better to continue growth while reproducing. Finally, we asked how resources can best be used for reproduction; about the relative merits of producing a few large young or many small ones; and of producing males or females.

6 Body temperature

This chapter is about how the temperatures of animals' bodies are affected by the environment and by the animal's metabolism, and how body temperature in turn affects the rates of physiological processes.

6.1 Temperature and activity

Enzyme-catalysed reactions are affected by temperature. If the temperature is low, reactions proceed slowly. Increasing the temperature increases the rate of the reaction, typically doubling it for every 10 K rise (see Box 6.1). Figure 6.1(a) shows, as an example, the rate of the reaction catalysed by citrate synthase from two species of moth. This is one of the enzymes of the tricarboxylic acid cycle that is so important in aerobic metabolism. The graph has a logarithmic vertical axis and is a straight line, showing that the relationship is exponential. Over the range of temperatures investigated, the rate is multiplied by a factor of 1.85 for every 10 K rise in temperature (but at higher temperatures the rate would fall as the enzyme was denatured).

Faster biochemical reactions imply faster metabolism. Figure 6.1(b) shows how the resting metabolic rate of a moth increases as its body temperature rises. As for individual reactions, the rate is roughly doubled for each 10 K rise over a wide range of temperatures. (In this graph neither axis is logarithmic, so the exponential relationship gives a curved line.)

Faster reactions enable muscles to exert more power. Figure 6.1(c) shows results

Box 6.1 **Temperature scales**

There is a slight awkwardness about scales of temperature which needs explanation. In the International System of units that scientists use, the kelvin (K) is the basic unit of temperature. On the Kelvin (absolute) scale of temperature, the freezing and boiling points of water are 273 K and 373 K, respectively. When discussing the temperatures of animals, it is generally more convenient to use the Celsius (centigrade) scale, on which water freezes at 0°C and boils at 100°C. Temperature differences have the same numerical values, whichever scale is used, so for them it seems best to call the unit the kelvin. It may seem anomalous to be told that the difference between 15°C and 10°C is 5 K, but this convention reduces the danger of mistaking a temperature difference for a temperature, and vice versa.

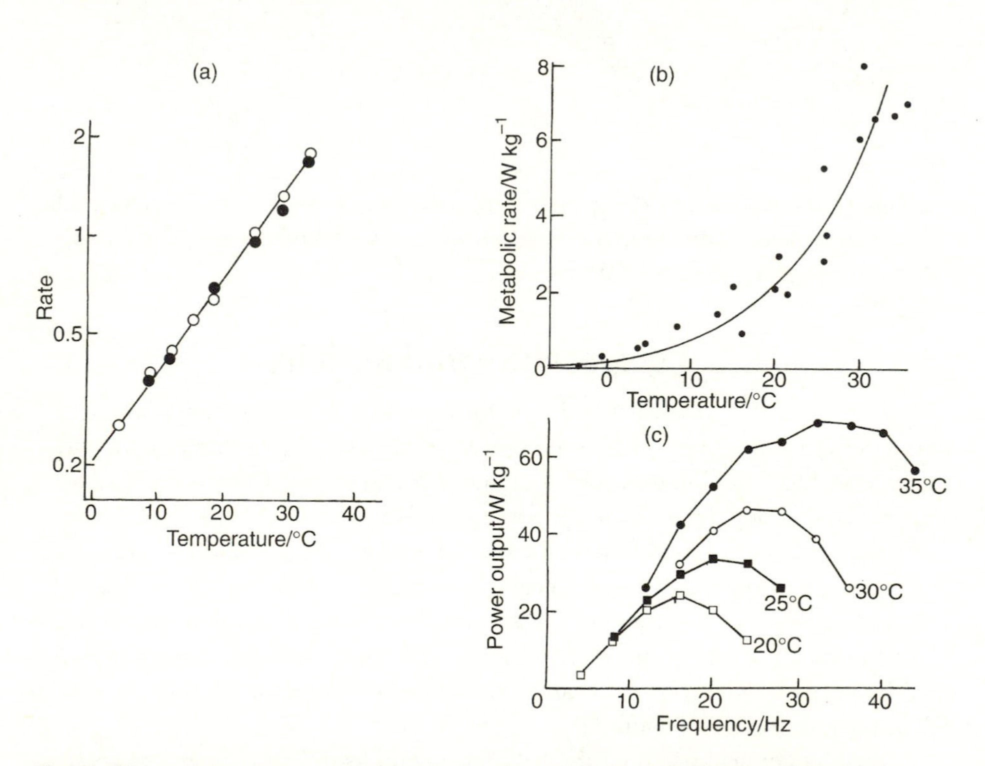

Fig. 6.1 Effects of temperature on the moths *Eupsilia* and *Manduca.* (a) The rate of the citrate synthase reaction for *Eupsilia* (filled circles) and *Manduca* (open circles). (b) The resting metabolic rate of *Eupsilia* and (c) the power output obtainable in work-loop experiments at different frequencies from *Manduca* flight muscle. ((a) redrawn from B. Heinrich and T. P. Mommsen (1985) *Science* **228**, 177–9; (b) from B. Heinrich (1987) *Journal of Experimental Biology* **127**, 313–32; and (c) from R. D. Stevenson and R. K. Josephson (1990) *Journal of Experimental Biology* **149**, 61–78.)

of work-loop experiments (see Section 4.4) on moth flight muscle. It shows the power obtainable at different frequencies of contraction, at each of four temperatures. As temperature increases, so does the optimum frequency that gives most power. The power output obtained at the optimum frequency doubles, for any 10 K temperature rise between 20 and 35°C. This is important for the moth; it cannot produce enough power to fly when its muscles are cooler than 29°C.

Increased power output means that (within limits) warm animals can move faster. Figure 6.2(a) shows the maximum sprinting speeds of lizards that were filmed as they ran along a 3 metre track. The higher the temperature, the faster in general they ran, but there is no perceptible increase in speed between 35 and 40°C, and the speed of another species that was tested in even hotter conditions fell slightly between 40 and 45°C. Speed is stride length multiplied by stride frequency. The graph shows that over most of the temperature range, stride length remained constant and the increase in speed was entirely due to increasing frequency.

Figure 6.2(b) shows sprinting speeds again, this time for goldfish startled by hitting their tank or dropping something in. The highest speeds were not at the highest temperatures. Notice the difference between fish that had been kept at different temperatures; each group swam fastest at or a little above the temperature to which it had been acclimated. Small differences were found in myosin from the muscles of the two groups.

Figure 6.2 refers to sprinting powered by anaerobic metabolism, but maximum aerobic metabolic rates also tend to rise with increasing temperature. Figure 6.3(a) shows resting metabolic rates of the lizard *Iguana* and the maximum rates that

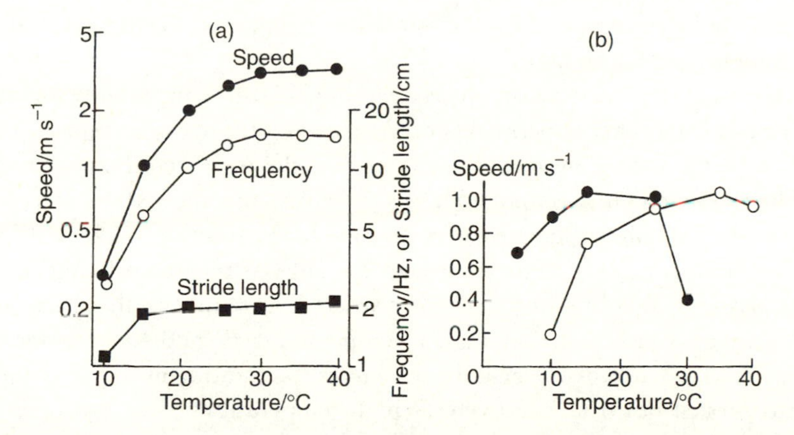

Fig. 6.2 The effects of temperature on sprinting speeds. (a) Maximum speed and corresponding stride lengths and stride frequencies plotted against temperature for the lizard *Sceloporus*; (b) peak speeds of startled goldfish (*Carassius auratus*) plotted against temperature. The fish had been kept at 10°C (filled circles) or 35°C (open circles)for the preceding 4 weeks. ((a) after R. L. Marsh and A. F. Bennett (1986) *Journal of Experimental Biology* **126**, 79–87; (b) after T. P. Johnston and A. F. Bennett (1995) *Journal of Experimental Biology* **198**, 2165–75.)

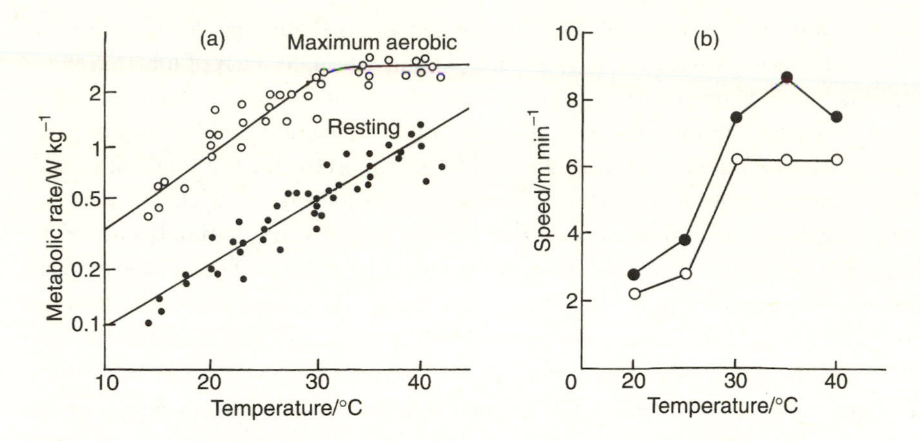

Fig. 6.3 Effects of temperature on metabolic rates and sustainable speed in the lizard *Iguana*. (a) Resting and maximum aerobic metabolic rates; (b) the maximum speeds that could be sustained for 10–20 min (filled circles) and 1 h (open circles), all plotted against temperature. ((a) after a figure and (b) from a table in W. R. Moberly (1968) *Comparative Biochemistry and Physiology* **27**, 1–20 and 21–32.)

could be obtained by running it on a treadmill. The resting rate increases by a factor of 2.2 for every 10 K rise in temperature, from 15 to 40°C. The maximum rate increases only up to 30–35°C, then levels off. The difference between the resting and maximum rates represents the energy available to power running by aerobic metabolism, and is greatest at 30–35°C, suggesting that the lizard should be able to sustain higher running speeds in this range of temperatures than in warmer or cooler conditions. Figure 6.3(b) shows maximum speeds sustainable for 10–20 minutes and for an hour.

Digestion as well as metabolism proceeds faster at higher temperatures. Figure 6.4 shows results from other experiments on *Iguana*. The lizards were kept in a cage heated by lamps, with a thermistor taped to their chests to record body temperature. They were fed on sweet potato (*Ipomoea*) leaves, and allowed to eat as much as they wanted. At higher temperatures they ate more, twice as much at 36° as at 30°C. The rate of passage of food through the gut was measured by including a strip of plastic in the food and observing when it appeared in the faeces; food passed through much faster at higher temperatures (Fig. 6.4(b)). Faster food passage would give no advantage if the food were less thoroughly digested, but the food was broken down as effectively at high temperatures as at low ones (Fig. 6.4(a)). This was demonstrated by measuring the dry masses both of the food that was eaten and of the faeces.

Thus as body temperature rises, within limits, we can expect to see faster metabolism, faster movement, faster feeding, and faster digestion. Higher temperatures generally increase the pace of life.

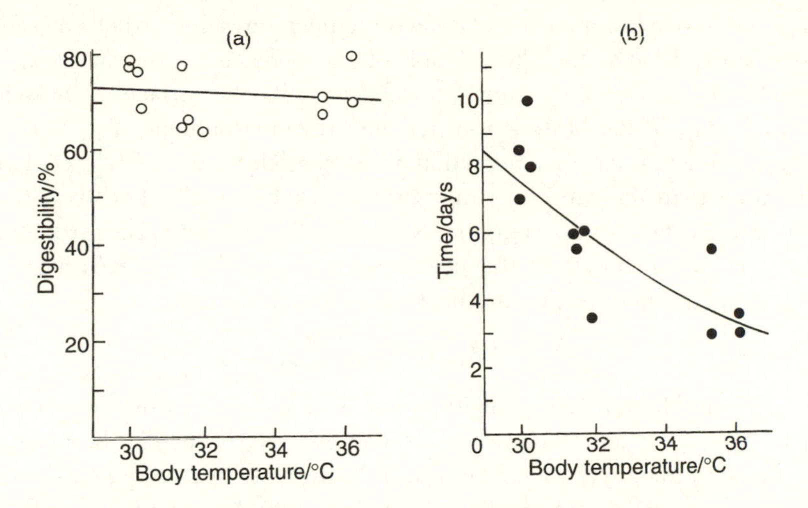

Fig. 6.4 The effects of temperature on *Iguana* lizards feeding on *Ipomoea* leaves. (a) The percentage of the food that was digested or fermented, calculated by comparing the dry masses of food and faeces (*Iguana* has microorganisms, in its hindgut, that break down cell wall materials). (b) Time taken for food to pass through the gut. (From W. D. van Marken Lichtenberg (1992) *Physiological Zoology* **65**, 649–73.)

6.2 **Heat exchange**

This section is about processes that affect the temperatures of animals and other bodies.

When heat is added to a body its temperature rises, and when it loses heat its temperature falls. Adding heat H to a body of mass m increases its temperature by ΔT

$$\Delta T = H/mC \qquad (6.1)$$

where C is the body's specific heat capacity. For water C is 4200 J kg^{-1} K^{-1}, but for animal tissues (which contain other substances as well as water) it is lower. Experiments on mouse carcasses gave a value of 3500 J kg^{-1} K^{-1}, which is probably typical.

When the sun starts shining on an animal such as a lizard, heat is added to the animal's body and the temperature rises. When a cool wind starts blowing, heat is removed from its body and its temperature falls. We need to understand how these and other influences affect the body temperatures of animals. Heat is transmitted from one body to another by three processes—conduction, convection, and radiation. We need to know a little about the physics of all three.

Figure 6.5(a) represents a section through an animal's body. The core, the inner parts of the body, must be at a uniform temperature because, if any part of it became hotter than others, the circulating blood would carry heat from there to the cooler parts and the temperature would be evened out. However, if the core of the body and the environment are at different temperatures there will be a gradient of temperature through the skin. Think of the body as a core at a uniform temperature T_{core} covered by skin of thickness s, with the surface of the skin at temperature T_{skin}. If the body is warmer than the environment, T_{core} is greater than T_{skin}, and heat is lost by conduction across the skin. If the body is cool, T_{core} may be lower than T_{skin} and heat may be conducted inwards. Let Φ_{cond} be the heat flux density through the skin, that is the heat flowing outwards through unit area of skin, in unit time. (If the heat flows inwards, Φ_{cond} is negative.) The subscript 'cond' reminds us that this heat travels by conduction.

$$\Phi_{cond} = (k_{skin}/s)(T_{core} - T_{skin}) \tag{6.2}$$

where k_{skin} is the thermal conductivity of the skin. We will discuss later how the conductivity depends on the nature of the skin, whether it includes blubber as in whales, fur as in mammals, or feathers as in birds. Note in the equation that the flux density is proportional to the temperature difference and inversely proportional to the thickness of the skin.

Heat may be removed from the outer surface of the skin by convection; that is, by cooler air (or water, in the case of an aquatic animal) flowing past. If there is nothing else to make the fluid (air or water) move, convection currents will develop. Fluid close to the body will be warmed by it and expand, so its density will fall and it will rise, carrying heat away. Its place close to the body will be taken by a fresh supply of cool fluid which, in turn, will be warmed and rise, removing more heat. This process is called free convection. Alternatively, a cool wind blowing past the animal or a cool water current flowing over it may carry heat away; this is called forced convection. For a dog indoors, free convection may often be more important than forced convection, but out of doors there is usually enough wind for forced convection to predominate.

Think of wind blowing over a stationary body. Even if the wind is strong, the air actually in contact with the body is stationary. There is a thin layer of air known as the boundary layer in which the air velocity increases from zero (at the

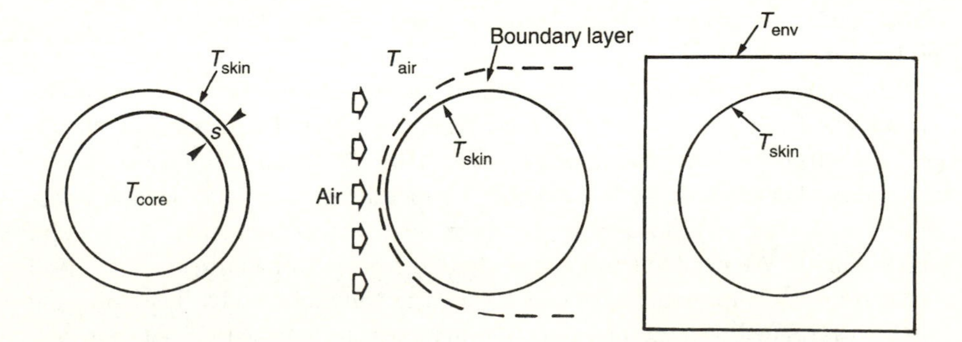

Fig. 6.5 Diagrams used to explain three processes of heat transfer. (a) Conduction, (b) convection, and (c) radiation.

body's surface) to almost the full speed of the unobstructed wind. Forced convection depends on heat being conducted through the slow-moving air of the boundary layer to the fast-moving air beyond. Consequently, the flux density can be calculated using a version of the conduction equation (eqn 6.2):

$$\Phi_{conv} = (k_{air}/\delta)(T_{skin} - T_{air}) \tag{6.3}$$

where k_{air} is the thermal conductivity of the air, T_{air} is its temperature, and δ is the effective thickness of the boundary layer. The boundary layer does not have a sharp outer edge and it is thinner on the windward side of the body than on the lee side (Fig. 6.2(b)), so the definition of its thickness has to be rather arbitrary. For a 6 mm diameter caterpillar in a gentle (1 m s^{-1}) wind it is about 0.6 mm. For a 600 mm diameter pig it is 4 mm in a 1 m s^{-1} wind and 1 mm in a 10 m s^{-1} one. The smaller the animal and the faster the wind, the thinner the boundary layer and the greater the convective flux density. The thicknesses in this paragraph have been calculated using an empirical equation that works reasonably well over this range of diameters and speeds:

$$\delta \approx 0.005d^{0.4}v^{-0.6} \tag{6.4}$$

where the thickness δ and the diameter d are in metres and the wind speed v in metres per second. In water the boundary layer would be five times thinner.

Heat may also be lost from the body by radiation. A surface whose absolute temperature is T_{skin} emits radiation with flux density σT_{skin}^4 where σ is the Stefan–Boltzmann constant, 5.7×10^{-8} W m^{-2} K^{-4}. However, it also receives radiation from its environment because surfaces in the environment at temperature T_{env} radiate with flux density σT_{env}^4. Generally, different parts of the environment are at different temperatures; vegetation may not be at the same temperature as bare ground, clouds are generally cooler than either, and the effective temperature of a clear sky is low because the upper layers of the atmosphere are very cold. For the present we will consider only the simple case in which all the objects around an animal are at the same temperature T_{env} (Fig. 6.2(c)). Then the net radiative heat flux density at an animal's surface (the flux density radiated by the animal minus the flux density received as incoming radiation) is

$$\Phi_{rad} = \sigma(T_{skin}^4 - T_{env}^4) \tag{6.5}$$

The absolute temperatures T_{skin} and T_{env} will commonly lie in the range 270 K (–3°C) to 310 K (37°C), so $T_{skin} - T_{env}$ will be small compared with T_{skin}. In that case we can use an approximation to eqn 6.5:

$$\Phi_{rad} \approx 4\sigma T_{skin}^3(T_{skin} - T_{env}) \tag{6.5}$$

Heat is conducted through an animal's skin, then convected or radiated from the skin's surface. When calculating the rate at which the animal loses heat, we have to take account of all three modes of heat transport. This is made easier by the equations for all three modes (eqns 6.2, 6.3, and 6.5) having the same form:

$$\Phi = (T_1 - T_2)/\mathcal{R} \tag{6.6}$$

where T_1 and T_2 are temperatures and $\mathcal{R}$ is a constant. This equation is very like Ohm's law of electric currents

current = (potential difference)/resistance

so we can think of $\mathcal{R}$ as a 'resistance'. (I have used quotation marks because $\mathcal{R}$ is not the same as the quantity that physicists mean when they refer to resistance to heat transfer.)

Because eqn 6.6 is so like the Ohm's law equation, we can calculate the rate of loss of heat from the body as if the heat were an electrical current flowing through a circuit involving three resistors (Fig. 6.6). The current representing the heat leaving the animal's body passes first through a resistance $\mathcal{R}_{cond}$ representing the resistance of the skin to heat flow by conduction, then through two resistances in parallel, $\mathcal{R}_{conv}$ and $\mathcal{R}_{rad}$, representing the resistances to the two alternative paths of heat loss from the skin surface, convection and radiation. The temperatures are represented by electrical potentials, T_{core}, T_{skin}, and $T_{air+env}$; I have assumed for simplicity that the air and everything else in the environment is at the same temperature $T_{air+env}$. Then the heat flux from the animal is

$$\Phi = (T_{core} - T_{air+env})/\mathcal{R}_{total} \tag{6.7a}$$

where the overall resistance is given by an equation like the one for electrical circuits

$$\mathcal{R}_{total} = \mathcal{R}_{cond} + \{1/[(1/\mathcal{R}_{conv}) + (1/\mathcal{R}_{rad})]\} \tag{6.7b}$$

and the resistances (obtained from eqns 6.2, 6.3, and 6.5) are

$$\mathcal{R}_{cond} = s/k_{skin} \tag{6.7c}$$

$$\mathcal{R}_{conv} = \delta/k_{air}$$

and

$$\mathcal{R}_{rad} = 1/4\sigma T_{skin}^{3}$$

Bigger animals have thicker skin than small ones. Geometrically similar animals would have skin thickness proportional to (body mass)$^{0.33}$ so $\mathcal{R}_{cond}$ would

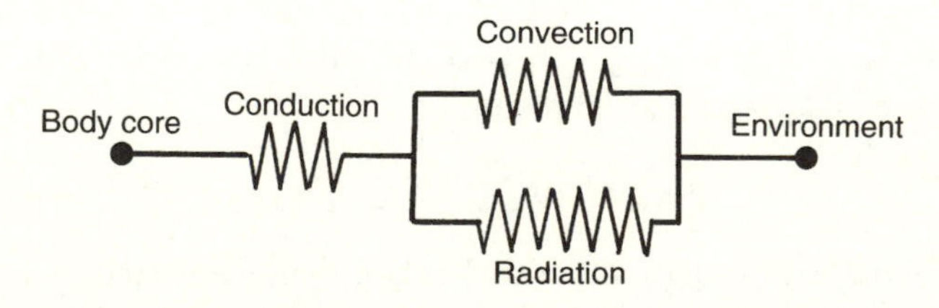

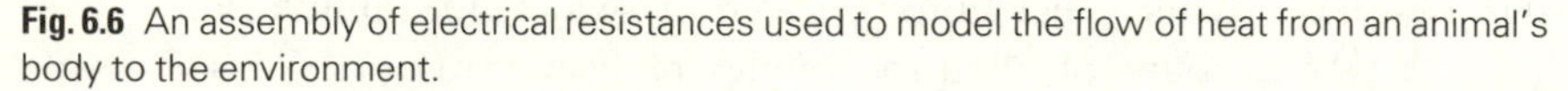
Fig. 6.6 An assembly of electrical resistances used to model the flow of heat from an animal's body to the environment.

also be proportional to (body mass)$^{0.33}$. $\mathcal{R}_{\mathrm{conv}}$ is also bigger for larger animals because boundary layer thickness is proportional to (body diameter)$^{0.4}$ (eqn 6.4). This means that for geometrically similar animals (diameter proportional to (body mass)$^{0.33}$), $\mathcal{R}_{\mathrm{conv}}$ is proportional to (body mass)$^{0.13}$. $\mathcal{R}_{\mathrm{rad}}$, however, is the same for animals of all sizes. This means that for very large animals the conductive resistance will be much larger than the others and the only one we have to worry about in rough calculations. For very small animals, however, it will be relatively small and can be ignored. Of the other two resistances, the convective resistance will be the lower for very small animals, so most of their heat will be convected away from the body surface, and the convective resistance will be the important one in calculations.

This may sound crazy. I have written that for small animals we can ignore the conductive resistance because it is small; and then that the convective resistance is important because it is smaller than the radiative resistance. The point is that if two resistances are in parallel, most of the heat will flow through the smaller of the two.

This section has told us that if an animal is large enough, its rate of heat loss in a cool environment will depend mainly on the resistance to heat conduction through its skin, fur, and so on; and that if the animal is small enough, the rate of heat loss will depend mainly on the convective resistance of the boundary layer. We will find out in the next section how 'large' is 'large enough' and how 'small' is 'small enough'.

6.3 Heating and cooling rates

The temperatures of small animals equilibrate quickly with the environment, but large animals heat and cool only slowly. We will be discussing heating and cooling rates of animals ranging from caterpillars to dinosaurs, but first we will consider some simple experiments that will help us to understand these rates.

The experiments were done with lizards and other reptiles, of a wide range of sizes, whose body temperatures were monitored (for example, by means of a thermocouple in the anus). In some of the experiments, the animals were left in a warm temperature-controlled chamber for long enough for their body temperatures to settle down at a constant value. Then they were moved to a cool chamber and their temperatures were observed as they cooled towards a new constant value. In other experiments they were moved from a cool chamber to a warm one and their temperatures were observed as they warmed up.

Consider an animal whose body temperature is initially T_1. At time zero it is moved to cooler surroundings at temperature T_2 and its temperature starts falling. At time t its temperature is T and it is losing heat with flux density $(T - T_2)/\mathcal{R}_{\mathrm{total}}$. This is the rate of heat loss per unit area; if the animal's surface area is A the rate of heat loss is $(T - T_2)A/\mathcal{R}_{\mathrm{total}}$. The animal's mass is m and its specific heat capacity is C, so removing a quantity H of heat from its body reduces its temperature by H/mC (eqn 6.1). Thus its rate of change of temperature is

$$dT/dt = -(T - T_2)A/mC\mathcal{R}_{total} \tag{6.8}$$

The temperature falls rapidly at first and then more slowly, as T approaches T_2. (Fig. 6.7 shows that a baked potato also cooled like this, when it was taken out of the oven.) The initial rate of fall of temperature is

$$(dT/dt)_{initial} = -(T_1 - T_2)A/mC\mathcal{R}_{total}$$

and if the temperature continued falling at this rate it would reach T_2 by time $mC\mathcal{R}_{total}/A$. This time is called the cooling time constant τ

$$\tau = mC\mathcal{R}_{total}/A \tag{6.9}$$

and is a useful indicator of the rate of cooling; animals with large time constants cool more slowly than those with small time constants. Figure 6.7 shows how the cooling time constant can be measured from a graph of temperature against time.

Figure 6.8(a) shows results from the experiments. Time constants range from 5 minutes for small (20 g) lizards to 2 hours for large (5 kg) ones. The graph is curved; we will try to explain why.

The graph has been drawn on logarithmic coordinates, so a straight line of gradient b would have the equation

$$\tau = am^b \tag{6.10}$$

(a is a constant.) We have seen that the time constant equals $mC\mathcal{R}_{total}/A$. Animals of all sizes are made of the same materials, so the specific heat capacity C can be assumed to be the same for them all. For geometrically similar animals, skin area A would be proportional to (body mass)$^{0.67}$ and we have seen that the conductive and convective resistances would be proportional to $m^{0.33}$ and $m^{0.13}$ respectively. If $\mathcal{R}_{cond}$ were the only important resistance to heat loss, we would expect the time

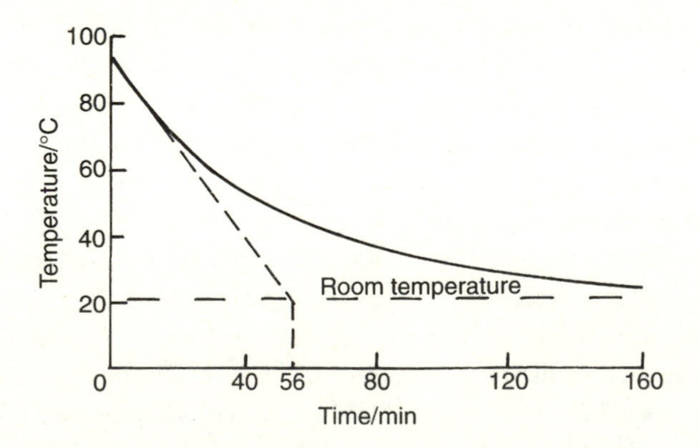

Fig. 6.7 A graph of temperature against time for an initially warm body in a cool environment (a baked potato, after removal from the oven). The cooling time constant was 56 min.

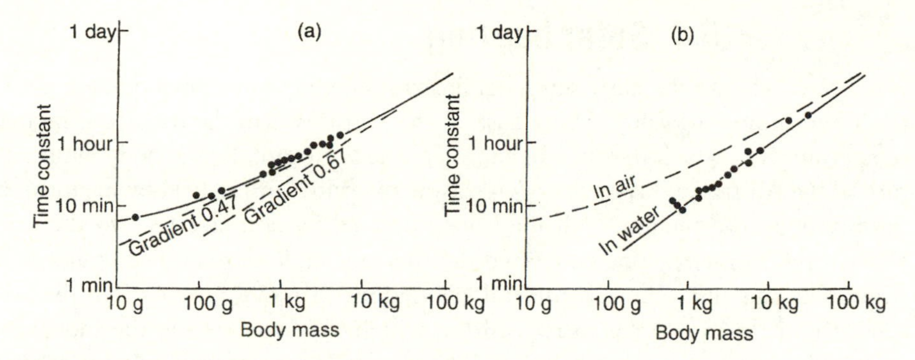

Fig. 6.8 (a) A graph of cooling time constant against body mass for reptiles in an air current of 3 m s^{-1}. Broken lines show gradients of 0.47 and 0.67. (b) The same curve with a similar graph for cooling in water. (Data are from papers listed by C. J. Bell (1980) *Journal of Experimental Biology* **86**, 79–85.)

constant to be proportional to $m \times m^{0.33}/m^{0.67} = m^{0.67}$ and the gradient of Fig. 6.8(a) to be 0.67. Similarly if $\mathcal{R}_{\text{conv}}$ were the only important resistance, we would expect the gradient to be 0.47. The graph is actually curved, with a gradient of about 0.4 (even lower than expected) for small reptiles and approaching 0.7 for large ones. This is reasonably consistent with our prediction in Section 6.2 that rates of heat loss would depend mainly on the convective resistance for small animals and on the conductive one for large animals.

Figure 6.8(b) shows time constants for cooling in water as well as in air. Water has 23 times the thermal conductivity of air and the boundary is five times thinner for the same rate of flow. For that reason, the convective resistance δ/k (eqn 6.7) will generally be much lower for an animal cooling in water than for a similar one cooling in air. However, the conductive resistance of the skin will be the same for both. The graph for cooling in water in Fig. 6.8(b) is very nearly a straight line of gradient 0.67, showing that the time constant depends mainly on the conductive resistance of the skin throughout the range of sizes of reptiles used in the experiments.

The largest animals used in the experiments were 30 kg crocodiles. You can get crocodiles a great deal bigger than that, but it takes a brave scientist to experiment with them. Figure 6.8(b) gives the impression that the graphs for cooling in air and in water would merge for animals of more than 100 kg. Small reptiles cool much faster in water than in air but large ones would cool almost equally fast in either fluid.

In other experiments, cool reptiles were warmed. The warming time constants were generally only about half as long as the cooling ones, probably because, in the range of temperatures used, the animals preferred to be warm. When being cooled, they probably reduced blood flow to the skin, increasing its effective thickness as a heat insulator; and while being warmed they probably increased blood flow to the skin, reducing the skin's effective thickness.

We will ask, in the next section, what these results mean for the lives of reptiles.

6.4 **Solar heating**

In the early morning, lizards generally have low body temperatures and move sluggishly. Many bask in the sun to warm themselves to higher temperatures at which they can be more active. For example, the body temperature of the Australian lizard *Amphibolurus* was measured in the field by means of a thermistor in the rectum, with the leads attached by adhesive tape to the tail. Light wires 50 metres long connected the thermocouple to recording equipment while allowing the animal considerable freedom to move around. The lizards spent the night in crevices in rocks, from which they emerged in the mornings with body temperatures around 25°C. They basked in the morning sun, warming at rates of around 1 K min^{-1} (1°C per minute) until their body temperatures were around 37°C, about the same as for humans and other mammals. Once warm they became active, but during the hottest part of the day they retired into the rock crevices and so avoided overheating. They emerged again as it got cooler in the evening, then returned to the crevices for the cold night. When in the sun they adjusted their positions to suit their requirements for temperature control. While basking to warm themselves they stood so as to expose a large area of skin to the sun's rays, but when getting too hot they either moved into the shade or faced directly into the sun so that the exposed area was as small as possible. Despite being 'cold blooded' reptiles, they kept their bodies at mammal-like temperatures for much of the day.

Small lizards have heating and cooling time constants of only a few minutes, so have only to bask for a few minutes to reach their preferred body temperature. The constants for large (1 tonne) crocodiles have not been measured, but extrapolation of the graphs in Fig. 6.5 indicates that they are of the order of 12–24 hours. Large crocodiles would take so long to heat or cool that they would never reach equilibrium in an environment with daily temperature fluctuations.

The sauropod dinosaur *Brachiosaurus* is estimated to have had an adult body mass of about 50 tonnes; some of the others seem to have been even heavier. Further extrapolation of the graphs in Fig. 6.8 indicates that 50 tonne dinosaurs would have had heating and cooling time constants of the order of a week. We cannot expect to be accurate when we extrapolate so far, from the animals of 30 kg or less that were used in the experiments to dinosaurs a thousand times as heavy, but it seems clear that such large reptiles must have had temperatures that remained almost constant day and night. They would have been warmer in summer than in winter, but a cold weekend would have made very little difference to their temperatures.

We can think about rates of heating in the sun in a different way. Think of a reptile that has reached equilibrium in the shade, then moves into the sun. The only disturbance to its heat balance is that it is now intercepting the sun's direct rays, which have radiant flux density Φ.

Suppose that the reptile stands with the long axis of its body at right angles to the sun's rays, so as to intercept as much energy as possible. If it were a cylinder of length l, diameter d, it would intercept the same amount of solar radiation as a rectangle of area ld, but the area of the cylinder (ignoring its ends) would be πld. The radiant flux density on the side of the cylinder directly facing the sun would be Φ, but the average radiant flux density over the whole of the cylinder's surface would be only Φ/π. Measurements of lizards' surface areas have shown that the area of a lizard of volume V is about $10V^{0.67}$, or $10(m/\rho)^{0.67}$ where m is the mass and ρ the density of the body. Thus the rate at which the reptile receives energy from the sun is $10(\Phi/\pi)(m/\rho)^{0.67}$. The rate $(\mathrm{d}T/\mathrm{d}t)_0$ at which its temperature starts rising is obtained by dividing this by its heat capacity mC, where C is the specific heat capacity of the body:

$$(\mathrm{d}T/\mathrm{d}t)_0 = 10\Phi/(\pi C m^{0.33}\rho^{0.67}) \tag{6.11}$$

Take $\Phi = 1000\ \mathrm{W\ m^{-2}}$ (Section 1.1); $C = 3500\ \mathrm{J\ kg^{-1}\ K^{-1}}$; and $\rho = 1000\ \mathrm{kg\ m^{-3}}$. Then for a 1 g reptile the initial rate of warming is $5\ \mathrm{K\ min^{-1}}$; for a 1 kg reptile it is $0.5\ \mathrm{K\ min^{-1}}$; and for a 1 tonne reptile it is $0.05\ \mathrm{K\ min^{-1}}$ or $3\ \mathrm{K\ h^{-1}}$. As before, we conclude that large reptiles heat up slowly.

Insects are small, in comparison to most reptiles, so will heat up quickly in the sun. Among them, 'furry' caterpillars, whose bodies are covered in stiff bristles, raise an interesting question. The bristles may serve one valuable function, by discouraging predators from eating the caterpillars, but have they a second function in heat balance? In laboratory experiments, intact furry caterpillars and ones whose bristles had been plucked out were heated by an artificial sun (a radiant electric heater) giving a radiant flux density of $700\ \mathrm{W\ m^{-2}}$. Thermocouples were used to monitor their body temperatures. Initially, with the 'sun' switched off, their temperatures were about 18°C. When the 'sun' was switched on, their temperatures rose. Twenty minutes later they had reached steady values of 23°C (plucked caterpillars) or 25°C (intact ones). Bristles have only a small effect on the temperature of a caterpillar in the sun, but a small increase in temperature will make a caterpillar more competitive by increasing the rate at which it can feed and grow. Apart from reducing the equilibrium temperature, plucking seemed to have no adverse effects on the caterpillars, which in due course pupated normally.

Can we explain the artificial 'sun's' effect on body temperature and the difference that the bristles make? Consider first a plucked caterpillar. If its long axis is perpendicular to the radiation, the mean radiant flux density over its whole body surface will be Φ/π, as for lizards. At equilibrium, this heat input is balanced by losses to the environment. We have already seen that, for sufficiently small animals, heat loss can be calculated reasonably accurately by taking account of convection, ignoring radiation and the conductive resistance of the skin. When convective heat loss, given by eqn 6.3, balances the gain from radiation

$$\Phi/\pi = (k_{\mathrm{air}}/\delta)(T_{\mathrm{skin}} - T_{\mathrm{air}})$$

$$T_{\mathrm{skin}} - T_{\mathrm{air}} = \Phi\delta/\pi k_{\mathrm{air}} \tag{6.12}$$

The radiant flux density was 700 W m^{-2}; the thermal conductivity of air is 0.025 W m^{-1} K^{-1}; and we saw in Section 6.2 that the boundary layer over a 6 mm diameter caterpillar in a 1 m s^{-1} wind (as in these experiments) would be about 0.6 mm thick. With these data, eqn 6.12 gives a temperature difference between body and air of 5 K. This matches the observed difference; the plucked caterpillars reached 23°C, and air temperature was 18°C.

The bristles on intact caterpillars hinder air movement close to the skin, so must thicken the boundary layer and increase the resistance to convective loss. It would not be easy to calculate how much thicker the boundary layer will become, but we should clearly expect the temperature to rise higher than for plucked caterpillars, as observed.

In our calculations about reptiles and caterpillars we have assumed that the animal absorbs all the solar radiation that falls on it. That is almost true for dark-coloured animals, but pale ones reflect a substantial fraction of the sun's rays. Many familiar beetles are black, but a black beetle in a hot desert is in danger of being heated by the sun to a lethal temperature. Some of the beetles of the Namib Desert have their black cuticle covered by waxy blooms in pastel shades. For example, *Zophosis* has a pink bloom which reflects about 25 per cent of solar radiation, much more than the 4 per cent that the cuticle reflects when the bloom has been cleaned off.

6.5 **Endotherms**

Most mammals maintain their bodies at 36–40°C and most birds at 40–43°C, even in environments in which a dead animal would freeze. They are enabled to do this by a combination of high metabolic rates and heat insulation: fur, feathers, or (in whales) blubber. The body temperatures of ectotherms such as the reptiles and caterpillars that we have been discussing are affected very little by their metabolism, but depend almost entirely on external influences. Those of endotherms (birds, mammals, and a few others that we will mention later) are greatly affected by heat from metabolism.

We saw in Section 2.4 that mammals and birds metabolize many times faster than reptiles of equal body mass. Their fur and feathers are good heat insulators because they trap a layer of stationary air over the surface of the body, preventing it from being convected away. Heat has to be conducted across this layer, so the conductive resistance to heat loss is greatly increased. Air and other gases have much lower thermal conductivities than liquids and solids. The thermal conductivity of air is 0.025 W m^{-1} K^{-1}. That of fur is a little higher, 0.04 W m^{-1} K^{-1}, because it includes solid hairs as well as air.

The blubber of whales, a layer of fat below the skin, is an alternative insulating layer. Its conductivity is about 0.24 W m^{-1} K^{-1}, six times as high as for fur, so a 6 cm thickness of blubber is needed to give the same insulation as 1 cm of fur. However, blubber has the advantage that it maintains its thickness when the

Table 6.1 Calculation of the expected temperature differences between whales and the surrounding water, using as examples a porpoise (*Phocaena phocaena*) and a blue whale (*Balaenoptera musculus*)

	Mass (kg)	Metabolic rate (W)	Surface area (m^2)	Thickness of blubber (cm)	Temperature difference (K)
Porpoise	60	72	1.0	2	6
Blue whale	120 000	23 000	180	10	53

The data are from J. E. I. Hokkanen (1990) *Journal of Theoretical Biology* **145**, 465–85.

whale dives. If whales were furry, the air in their fur would be compressed by the increased pressure when they dived, and the deeper they went the less effective the fur would be.

Table 6.1 shows how much effect the blubber can be expected to have, for whales of different sizes. The metabolic rates are minimal values, estimated using eqn 2.5. Little is known about whale metabolic rates, because of the practical difficulties of measuring them, but what data we have indicate that the equation works reasonably well for whales.

When body temperature is constant, the metabolic rate R balances the rate of heat loss through the blubber:

$$R = \Phi_{cond} A_{skin} = (A_{skin} k_{skin}/s)(T_{core} - T_{skin})$$

where A_{skin} is the area of the skin and Φ_{cond} is the flux density of conducted heat, as given by eqn 6.2. By rearranging the equation we get the temperature difference across the skin, $T_{core} - T_{skin}$

$$T_{core} - T_{skin} = Rs/A_{skin} k_{skin} \qquad (6.13)$$

We have seen that the thermal conductivity k_{skin} of blubber is about 0.24 W m^{-1} K^{-1}. The temperature differences shown in Table 6.1 have been calculated from the metabolic rates, areas, and blubber thicknesses given in the preceding columns, using eqn 6.13. The temperature difference is much larger for the larger whale, which is as we should expect: geometrically similar whales would have blubber thickness and area proportional to (mass)$^{0.33}$ and (mass)$^{0.67}$ respectively, and metabolic rates of mammals are generally proportional to (mass)$^{0.76}$, so eqn 6.13 gives temperature differences proportional to mass to the power $0.76 + 0.33 - 0.67 = 0.42$.

The temperature differences we have calculated are differences across the blubber, between the core of the body and the skin surface. What we would really like to know is the difference between the core of the body and water far enough from the animal not to be warmed by it. Conveniently, these temperature differences are almost identical because there is very little temperature difference across the boundary layer. This follows from what we have already noted; for large animals, especially in water, the resistance to radiative and convective heat loss from the skin surface is small enough to be ignored.

Table 6.1 tells us that the minimal metabolic rate of a 60 kg porpoise is enough to maintain the animal's body temperature of 37°C in water 6 degrees colder, at 31°C. In water colder than that, the animal would have to raise its metabolic rate.

The table also tells us that the minimum metabolic rate of a 120 tonne blue whale is enough to overheat the animal, raising its temperature well over 40°C even when the sea water is at its freezing point, –2°C. If the whale is active, metabolizing at more than the minimal rate, the tendency to overheat will be exacerbated.

Overheating is avoided by reducing the insulating effect of the blubber. The whole thickness of the blubber functions as insulation only if no blood is allowed to flow through its blood vessels. When heat needs to be lost, blood is allowed to flow through vessels in the blubber. This is apparently very effective; though blue whales spend the southern summer feeding in the cold waters of the Antarctic, they move north into warmer waters for the winter and even there do not overheat.

We will return in due course to aquatic animals, to discuss fishes that keep parts of their bodies well above the temperature of the water. First, however, we will consider the heat balance of terrestrial mammals that depend on fur rather than blubber for insulation, and of birds that depend on feathers.

Figure 6.9 shows results from an experiment in which birds were kept at different temperatures and their metabolic rates were measured. The metabolic rate of this particular species was lowest, and constant, between about 18 and 33°C; this range is called the thermoneutral zone. Below 18°C the metabolic rate was higher, rising (as temperature fell) roughly in proportion to the difference between the bird's body temperature of 40°C and the environmental temperature. Above 33°C the temperature also rose. The interpretation of this graph is that the metabolic rate was **either** whatever was needed to perform essential functions such as protein turnover and ion pumping (see Chapter 2) **or** sufficient to maintain body temperature, whichever was the greater. In the thermoneutral zone the metabolism needed for essential functions was enough or more than enough to replenish heat lost to the environment. Any surplus heat was dissipated by flattening the feathers against the body (thinning the heat insulation) or by directing more blood to the uninsulated legs. Below 18°C the heat produced by minimal metabolism was not enough to replace heat losses, so the metabolic rate had to be increased. Above 33°C, the birds panted. Their breathing muscles did work to ventilate the lungs very rapidly so the metabolic rate was elevated, but the rapid ventilation caused rapid evaporation of water which was lost as vapour in the breath (the cooling effect of evaporation is discussed below). The extra heat loss due to panting was greater than the increase in the metabolic rate, so the net effect was heat loss.

With these results in mind, look at Fig. 6.10. The continuous lines and filled symbols show minimal metabolic rates, observed in the thermoneutral zone. The

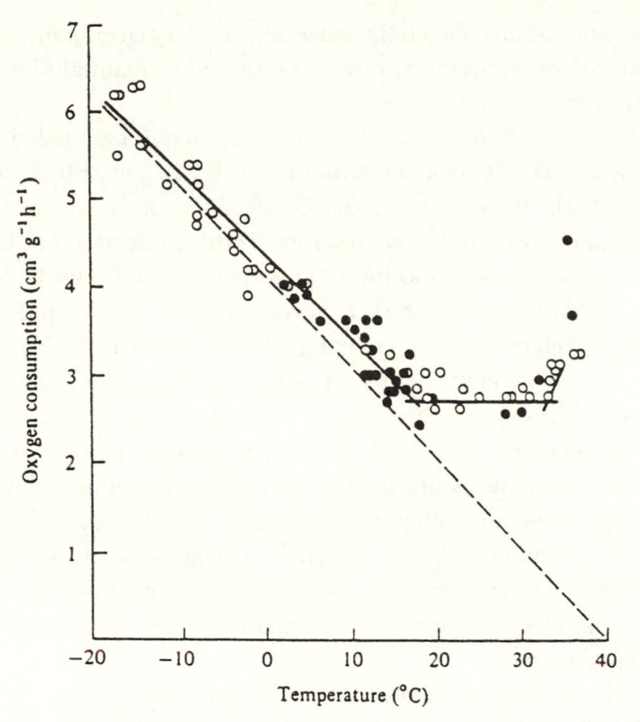

Fig. 6.9 A graph of metabolic rate against environmental temperature for cardinals (*Richmondena cardinalis*), a bird of about 40 g body mass. A rate of oxygen consumption of 1 $cm^3\ g^{-1}\ h^{-1}$ corresponds to a metabolic rate of 5.6 W kg^{-1}. Filled circles, birds kept indoors; open circles, birds kept out of doors in cold weather. (From W. R. Dawson (1958) *Physiological Zoology* **31**, 37–48. Copyright 1958 by the University of Chicago.)

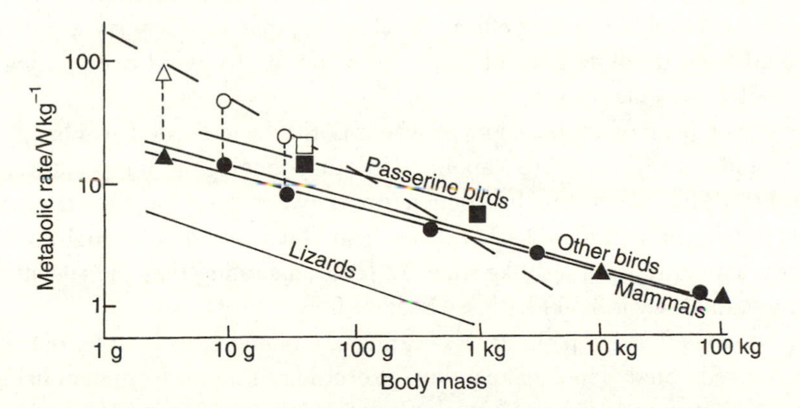

Fig. 6.10 Graphs of resting metabolic rates against body mass for various lizards, birds, and mammals. Filled symbols show resting rates in the thermoneutral zone for passerine birds (squares), non-passerine birds (triangles), and mammals (circles). Open symbols show rates for some of the same species at 10°C. The regression lines summarize data from 35 passeriform and 58 other bird species; 13 species of mammal; and 13 species of lizard at 37°C. (After R. McN. Alexander (1975) *The Chordates*, Cambridge University Press, where sources of data are listed.)

open symbols show resting metabolic rates in a cool environment, at 10°C. All these are greater than the minimal rates, as required to maintain body temperature below the thermoneutral zone.

The broken line shows the metabolic rates estimated to be needed to maintain a body temperature of 37°C in an environment at 10°C. Very simple assumptions were made to obtain these. The surface area of the body was assumed to be $10(\text{volume})^{0.67}$, an expression that we used for lizards in Section 6.4 which is also about right for humans, cows, and frogs. The thickness of the fur or feathers was assumed to be $0.3(\text{volume})^{0.33}$, which is about right for shrews and Arctic foxes (mammals with a relatively thick covering of fur) and would also be right for geometrically similar animals of different sizes, with the same ratio of insulation thickness to body diameter. The thermal conductivity of the insulating layer of fur or feathers was assumed to be 0.04 W m^{-1} K^{-1}, as measured for fur. It was assumed that there was no wind, so that only free convection occurred (so the equation for convection was different from eqn 6.3). This, and the assumption that the air and everything else in the environment were at 10°C, match the conditions in the laboratory experiments in which the metabolic rates shown by the open symbols were measured. The theoretical line agrees well with the measured metabolic rates.

Notice that the broken line showing heat loss at 10°C crosses the continuous lines showing metabolic rates in the thermoneutral zone, at moderate body masses. A 150 g passeriform bird or a 1 kg mammal would enter the thermoneutral zone at 10°C if they were as well insulated as a shrew or an Arctic fox. In environments warmer than 10°C they would have to thin their insulation by flattening their fur or feathers against the body, or dissipate excess heat by panting or sweating. Even when flattened as much as possible, a thick coat of fur would have a considerable insulating effect. An elephant that was as furry as an Arctic fox would have to allow a lot of water to evaporate to avoid overheating in a tropical environment.

The latent heat of evaporation of water is the heat needed to change unit mass of water from liquid to vapour, without changing its temperature. It is 2.4 MJ kg^{-1}. That means that by losing a litre of water (1 kg) by panting or sweating, an animal can get rid of 2.4 MJ excess heat. That would be enough to reduce the body temperature of a 70 kg man by 10 K (assuming that the specific heat capacity of the body is 3.5 kJ kg^{-1}; see Section 6.2).

The cooling effects of panting or sweating are often very valuable, but should not be overestimated. Think of an animal exercising strenuously, metabolizing fat, with no opportunity to drink; for example, a bird migrating over a desert or sea. The enthalpy of combustion of fat is −39 MJ kg^{-1} (Table 2.1). Muscles working with 25 per cent efficiency will release 75 per cent of this energy as heat, about 30 MJ for every kilogram of fat that is metabolized. Metabolism of a kilogram of fat also produces about a kilogram of water (this can be worked out from eqn 2.1). If the water so produced is allowed to evaporate it will remove 2.4 MJ as latent heat, but this is only 8 per cent of the heat produced by the metabolism. Some additional heat can be shed if the animal is prepared to allow the water content of its body to fall. For example, in one experiment the water vapour lost in the breath of budgerigars flying in a wind tunnel accounted for 15 per cent of the heat they were producing. Only half of the water can have come from the metabolized fat, and the net loss of water must have been about equal in mass to the fat used. A migrating bird with no opportunity to drink simply could not afford to lose water at this rate during a long flight in which it used most of its fat reserves. For example, yellow wagtails (*Motacilla flava*) cross the Sahara on their migration from Nigeria to Europe. Birds leaving Nigeria were found to have a mean mass of 24 g, including 7.4 g fat. Those arriving north of the Sahara had a mean mass of only 15 g, having lost nearly all their fat and some of their water.

It is only on small birds and mammals and on ones that live in cold climates (like the Arctic fox) that we find insulation as thick as was assumed in the theoretical calculations (Fig. 6.10). The largest land mammals, elephants and rhinoceros, have no fur and are insulated only by their skin (but woolly mammoths and rhinoceros flourished in the cold climate of the Ice Age).

Figure 6.10 shows that a 2 g hummingbird or shrew in an environment at 10°C would have to metabolize at five times its minimal rate. This temperature is cool, but not really cold. The highest metabolic rates that birds and mammals can maintain for long periods seem never to exceed seven times the minimal rate, as will be discussed in Section 7.10. The smallest shrews and hummingbirds have masses of about 2 g and may be close to the lower limit of size for animals that maintain body temperatures of around 40°C in an environment that is sometimes cool.

Such small endotherms can make useful savings of energy by allowing their body temperatures to fall when they are inactive, going into a state of torpor. Hummingbirds are active by day and become torpid on cold nights. Bats that are active at night become torpid at their daytime roosts. And larger mammals such as hedgehogs (*Erinaceus*, 0.5–1.0 kg) allow their body temperatures to fall while they are hibernating. Substantial savings of energy are possible only if the period of torpor is fairly long, compared to the animal's cooling time constant. If an elephant dropped its metabolic rate at night it would cool only a little before morning and its heat loss would be reduced only slightly.

As well as birds and mammals, a few other animals have permanently elevated body temperatures. Adult leatherback turtles (*Dermochelys*) have masses around 400 kg. The body temperatures of freshly caught specimens have been measured and found to be up to 18 K warmer than the water in which they had been swimming (for example, a body temperature of 26°C in water at 8°C). A typical 400 kg reptile at 30°C would be expected to have a resting metabolic rate of about 30 W and a 400 kg mammal would have a minimal metabolic rate of 300 W. The actual metabolic rate of *Dermochelys* is between these rates, about 150 W.

There are also endothermic fishes—tunas and a few species of shark. In some tunas, parts of the swimming muscle may be as much as 20 K above the

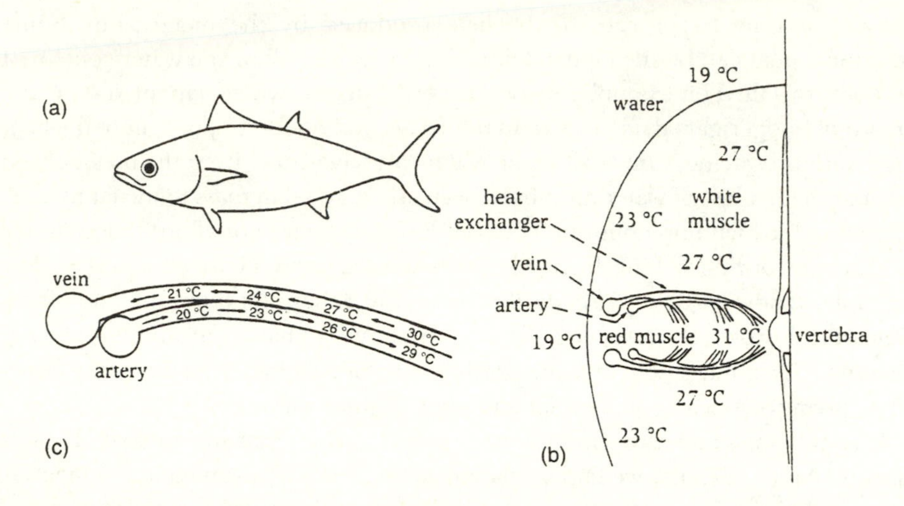

Fig. 6.11 (a) A sketch of a tuna and (b) a diagrammatic transverse section showing the red muscle and its blood supply. The temperatures were measured in a freshly caught bluefin tuna (*Thunnus thynnus*) which had been swimming in water at 19°C. (c) A diagram illustrating the principle of the heat exchanger, showing possible (not measured) temperatures. (From R. McN. Alexander (1990) *Animals*, Cambridge University Press.)

temperature of the water. These active fishes, which swim continuously, have an unusually large proportion of red (aerobic) muscle. In most fish, the aerobic swimming muscle forms a thin layer close under the skin, but in tunas it extends deep into the body (Fig. 6.11(b)). Temperatures within the muscle have been measured by probing freshly caught tunas with long needles with thermistors at their ends. Generally, all the muscle is warmer than the water that the fish had been swimming in, but the deep parts of the red muscle are the warmest. Figure 6.11(b) shows data from a typical individual fish.

Temperatures like these could not be maintained by a fish with a conventionally arranged blood system, because heat loss from the gills would prevent body temperature from rising. Suppose that there is a temperature difference ΔT between the blood flowing through the gills and the water flowing over them (the blood is warmer, because it has come from the warm muscles). Suppose also that the partial pressures of oxygen in the blood and water differ by ΔP (higher in the water than in the blood, whose oxygen content has been depleted in its passage through the tissues). Heat is conducted from blood to water with flux density Φ_{heat}

$$\Phi_{\text{heat}} = (k/s)\Delta T$$

where k is the thermal conductivity and s is the effective thickness of the barrier between the blood and the water. Oxygen diffuses from blood to water with flux density Φ_{oxygen}

$$\Phi_{\text{oxygen}} = (D/s)\Delta P$$

where D is the diffusion constant. By dividing the first of these equations by the second we get

$$\Phi_{heat}/\Phi_{oxygen} = k\Delta T/D\Delta P$$

$$\Delta T = (D\Delta P/k)(\Phi_{heat}/\Phi_{oxygen}) \qquad (6.14)$$

Now we will put numbers into the equation. The diffusion constant D for oxygen diffusing through water is 6×10^{-11} m^2 atm^{-1} s^{-1}. The thermal conductivity k of water is 0.6 W m^{-1} K^{-1}. The partial pressure difference ΔP cannot be greater than 0.2 atm, the partial pressure of oxygen in the atmosphere or in air-saturated water. Metabolism using a litre of oxygen makes 20 000 J of energy available (Section 2.3), so metabolism using a cubic metre makes 2×10^7 J available. The heat lost must come from metabolism, so $\Phi_{heat}/\Phi_{oxygen}$ cannot exceed 2×10^7 J m^{-3}. Putting all these values in eqn 6.14, we find that ΔT cannot exceed 0.0004 K. The fish cannot be appreciably warmer than the water, unless of course it has recently moved from warm to cooler water.

That argument overstates the case; measurements on *Tilapia*, a teleost with conventional blood circulation, have shown that the muscles are generally about 0.4 K warmer than the water. This is probably because the arteries and veins of the swimming muscles run quite close together, allowing some heat exchange of the kind that (as we shall see) makes the high temperatures of tuna possible. However, the temperature difference in *Tilapia* is tiny, compared to the large differences found in tuna.

In typical teleosts, the arteries to the swimming muscles are branches of the dorsal aorta, and the veins are branches of the posterior vena cava. The aorta and the vena cava run longitudinally along the fish, immediately ventral to the vertebral column. In contrast, the muscles of tuna are served by arteries and veins that run close under the skin, where the temperature is much lower (Fig. 6.11(b)). Fine branches of these arteries and veins run into the muscle, with the venous branches parallel to the arterial ones, and in close contact. The blood in the arteries under the skin is cool, close to the temperature of the water. The blood leaving the warm muscles is much warmer. The warm blood in the fine branches of the veins loses heat to the cooler blood flowing close to it, in the fine branches of the arteries. If these parallel blood vessels are long enough, the venous blood will be cooled almost to the temperature of the water, before reaching the veins under the skin. At every point on a venous branch, the blood must be slightly warmer than in the adjacent arterial branch (otherwise no heat would be exchanged) but Fig. 6.11(c) shows how the venous blood may be cooled almost to the temperature at which blood arrives in the arteries. The same principle is used in industry, in countercurrent heat exchangers, to recover heat from fluids leaving a warm reactor. These exchangers consist of bundles of parallel pipes, in which adjacent pipes carry fluid in opposite directions.

There must be some diffusion of oxygen from the arterial to the venous blood, in the tuna's heat exchanger, as well as conduction of heat in the opposite direction. However, the ratio of the thermal conductivity to the diffusion constant for oxygen ensures that this effect is very slight. The system is an effective heat exchanger but (fortunately for the fish) a very ineffective oxygen exchanger.

The warmest muscle is the deep part of the red muscle. Its high temperature makes higher power outputs possible, than if it were cool, enabling the fish to sustain faster swimming. This has been demonstrated by work-loop experiments (see Section 4.4) on bundles of muscle fibres taken from yellowfin tuna (*Thunnus albacares*). As for insect muscle (Fig. 6.1(c)), optimum cycle frequency and maximum power output both increased rapidly with increasing temperature. Between 20 and 30°C, the optimum frequency doubled, from 4 to 8 Hz, and maximum power increased by a factor of 1.9. In a large aquarium, the fish spent most of their time swimming with tail beat frequencies of less than 2 Hz, but used frequencies of 8–12 Hz in short bursts of speed when chasing prey.

Bundles of deep red muscle fibres worked well at 30°C but deteriorated at 35°C. In contrast, similar preparations from bonito (*Sarda chiliensis*, an ectothermic member of the same family) deteriorated even at 30°C. Tuna red muscle seems to be adapted to work at high temperatures. Another remarkable feature of tunas is that their gill areas are greater than expected for fish of their size, making possible the high rates of oxygen uptake that are needed to supply the large, warm red muscles.

6.6 Temporary endotherms

Bumble-bees are fairly large insects, ranging from about 60 mg for a worker of a small species to 700 mg for a queen of a large one. They have a fur-like, heat-insulating coat on their bodies. Like other insects, they are capable of extremely high metabolic rates when flying. These characteristics enable bumble-bees (and many other insects, such as large moths) to be endothermic while they are active.

To see how this is possible, look back at Fig. 6.10. The broken line shows the metabolic rates calculated to be needed to keep mammals' bodies 27 K warmer than the environment. Extrapolation of the line to lower body masses indicates that a 0.1 g (bumble-bee-sized) mammal would need a metabolic rate of about 600 W kg^{-1}. The metabolic rates of flying bumble-bees are about 350 W kg^{-1}, which suggests that they should be able to keep themselves $27 \times 350/600 = 16$ K warmer than the environment. That is a crude argument (bumble-bees are not tiny mammals) and it underestimates the temperature differences that are possible because it assumes wrongly that the whole body is heated.

Bernd Heinrich, who is responsible for much of what we know about endothermic insects, measured the temperatures of bumble-bees foraging in the wild in Maine. He grabbed them with a gloved hand as they fed on flowers, then jabbed a thermistor probe into them. That way he was able to get a temperature reading within four seconds of grabbing the bee. He found that the temperature of the

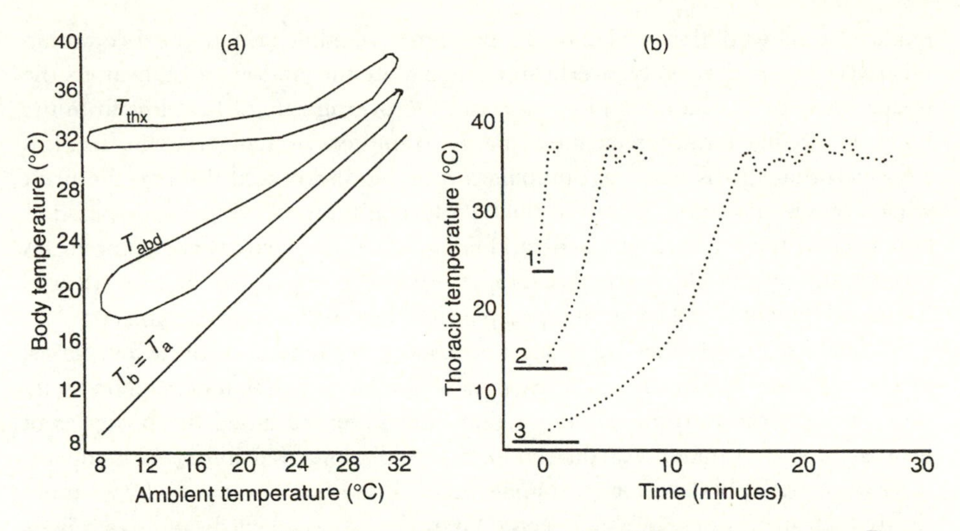

Fig. 6.12 (a) Temperatures of foraging bumble-bees (*Bombus*) at different air temperatures. The polygons enclose the thorax temperatures T_{thx} and abdomen temperatures T_{abd} observed at different ambient temperatures. The line of equality between body and ambient temperatures ($T_b = T_a$) is also shown. (b) Records of the temperatures of the thoraces of bumble-bees, warming themselves from different initial temperatures. (By permission of the publishers from B. Heinrich (1993) *The Hot-blooded Insects* Harvard University Press, Cambridge, Ma. Copyright 1993 by B. Heinrich, and B. Heinrich (1975) *Journal of Comparative Physiology* **96**, 155–66. Copyright 1975 by Springer-Verlag.)

thorax was about 33°C, when the air temperature was anywhere in the range from 9 to 24°C. On hotter days, especially in the sun, thorax temperatures might rise to 40°C. The abdomen was cooler, and its temperature was affected more by changes of air temperature (Fig. 6.12(a)). Thus active bumble-bees keep their thoraces at temperatures that approach typical mammalian body temperatures.

Resting bumble-bees allow their bodies to cool to air temperature. While they are cool, they cannot fly; the minimum thorax temperature for flight is about 29°C. That means they have to warm themselves up in preparation for flight. They do this by a process very like shivering—the flight muscles become active with a high metabolic rate, but the resulting movements are small. Figure 6.12(b) shows records of warming from different initial temperatures, obtained from thermocouples in the thoraces of bees in a laboratory. Notice how the rate of warming accelerates as the bee's temperature rises. Warmer muscles are capable of a higher metabolic rate, so can warm themselves faster.

Blood circulates all around a bee's body, but the thorax is nevertheless kept warmer than the abdomen. As in tunas, a countercurrent heat exchanger maintains the temperature difference. This is located in the narrow waist that connects thorax and abdomen. (Strictly speaking, the waist is between the first and second abdominal segments, so what I am for convenience calling the thorax actually includes one segment of the true abdomen.) The body cavity is a blood-filled haemocoel. Blood is carried forward through the waist in the aorta, but makes its way back in the haemocoel, and because the waist is so narrow heat is readily exchanged there (Fig. 6.13). In warm weather, when the bee would otherwise be in danger of overheating, it allows the abdomen to heat to the temperature of the thorax and so increases the overall rate of heat loss from the body. To do this, it must make the countercurrent heat exchanger ineffective. It is suspected that this is done by pumping a little blood forward through the waist while none is allowed to return in the haemocoel, then allowing a little blood to flow back in the haemocoel, to the abdomen, while the heart is not pumping. A mechanism that could make this happen has been suggested. If forward and backward flow of blood occur at separate times, heat will not be exchanged.

We started this chapter by seeing how increasing temperature (within limits) increases the rates of biochemical reactions and of processes that depend on them, such as digestion and muscle contraction. Then we examined the processes of heat exchange (conduction, convection, and radiation) that affect the temperatures of animals' bodies. Large animals such as dinosaurs warm and cool more slowly than small ones such as insects. We discussed how lizards and caterpillars

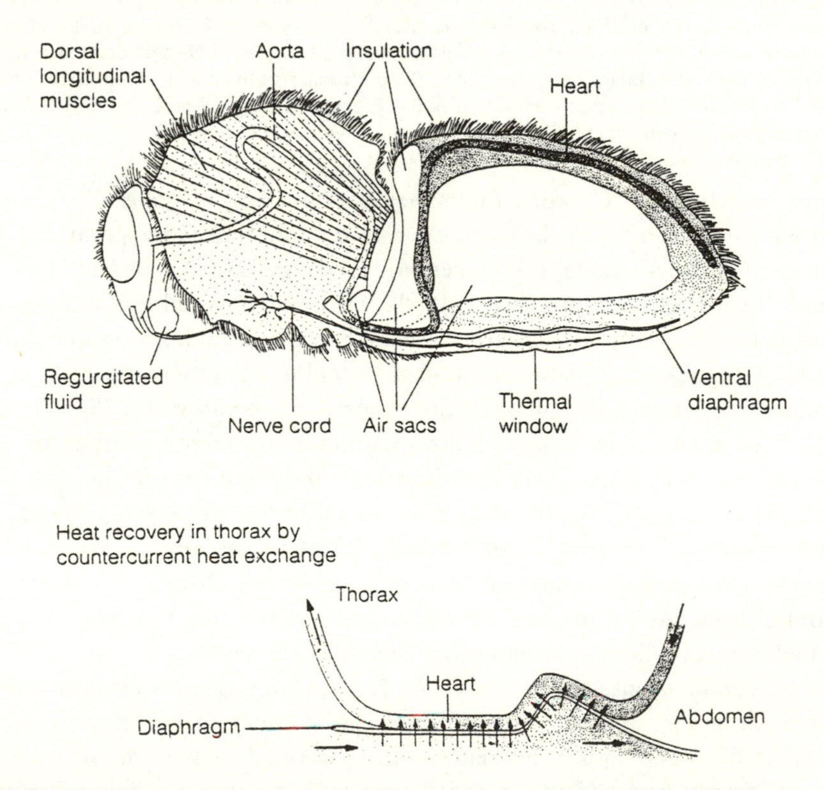

Fig. 6.13 Blood circulates between thorax and abdomen in a bumble-bee, forward in the aorta and back in the haemocoel. In the narrow waist, heat flows from the warm blood (light stipple) travelling back from the thorax, to the cool blood (dark stipple) travelling forward from the abdomen. An air sac at the anterior end of the abdomen serves as a heat insulator, resisting conduction of heat from thorax to abdomen where their walls are in contact. (From B. Heinrich (1976) *Journal of Experimental Biology* **64**, 561–85.)

take advantage of solar radiation to warm them, enabling them to be more active or to feed and grow faster. Finally, we discussed the endotherms, the animals that rely on their own metabolism to maintain high body temperatures. We saw that small endotherms such as shrews require very high metabolic rates, to keep themselves warm in cold environments. As well as the birds and mammals, some large fishes are endotherms, relying on countercurrent heat exchangers to prevent excessive heat loss through the gills. And many insects are temporary endotherms, warming their bodies to mammal-like temperatures whenever they fly.

7 Energy budgets

Previous chapters have told us how much energy is needed for particular functions such as growth, maintaining body temperature, or running at some particular speed. Now we ask how much energy animals use in their everyday lives, and how it is divided between functions.

7.1 Field metabolic rates

We have seen how animals' rates of energy consumption have been measured in laboratories, generally by measuring oxygen consumption or heat production, but now we want to know about the energy consumption of animals that are ranging freely in their natural environments. For this purpose, the doubly labelled water technique has been especially useful. It enables biologists to find out how much oxygen an animal has used between two occasions on which it is captured, without attaching any apparatus to it. The animal is given a harmless injection and released. When it is recaptured, a sample of its blood or other body fluids is taken for analysis.

The animal is caught and given an injection of water labelled with isotopes both of hydrogen and of oxygen. Alternatively (as is usual in experiments on humans) the labelled water can be given by mouth. It contains proportions both of $^{2}H_2O$ and of $H_2{}^{18}O$. The ^{2}H becomes mixed throughout the water in the body. The ^{18}O also gets mixed all through the water in the body, but in addition gets mixed throughout the body's bicarbonate and carbon dioxide by the reaction

$$H_2O + CO_2 \Leftrightarrow H^+ + HCO_3^-$$

Any carbon dioxide released from the body carries some of the ^{18}O away with it, but ^{18}O is also lost in any water that escapes from the body as urine, sweat, or vapour. Thus the rate at which ^{18}O disappears from the body tells us the total rate at which oxygen is being lost as water or carbon dioxide. The rate at which ^{2}H disappears tells how fast hydrogen is being lost as water. If we know both these things, we can calculate how fast the body is producing carbon dioxide. From that the metabolic rate can be calculated. Aerobic metabolism producing a cubic centimetre of carbon dioxide releases 21 joules if carbohydrate is being oxidized, and 28 joules in the case of fat. This can be worked out in the same way as was done for oxygen, in Section 2.3.

After the initial dose of labelled water has had time to distribute itself around the body, a sample of a body fluid (blood, urine, or saliva) may be taken. Alternatively, if the dose is precisely known, this may not be necessary. The animal is released for a period of hours or days, then recaptured so that another sample of body fluid can be taken. The ^{2}H and ^{18}O contents of the samples are measured by mass spectrometry and the rate of loss of carbon dioxide is calculated.

The method was originally devised for experiments with small mammals. It has been used for animals up to the size of deer and seals but is expensive for large animals, because they need large doses of the isotopes.

Field metabolic rates of some mammals and birds, measured by the doubly labelled water technique (except in the cases of the humans) are shown in Fig. 7.1. The data refer to a 24 h day in the natural habitat. As a general rule, field metabolic rates are around three times the minimal rates. These are 24 h mean rates, so if the animal spends part of each day sleeping or inactive, with a low metabolic rate, the rate must be higher while it is active.

The two humans are a labourer (with the higher metabolic rate) and an office worker. The graph makes them seem rather inactive in comparison to the animals, but the animals of human-like size happen to be seals and deer, members of groups in which minimal metabolic rates, and possibly also field metabolic rates, tend to be higher than average.

The field metabolic rates of some iguanid lizards have also been measured by the doubly labelled water technique. They were found to be only 6–7 per cent of the rates for mammals of equal mass. Regrettably, good information about field metabolic rates of invertebrates is sparse.

In the following sections we will ask what animals do in their natural habitats and how they use their energy. This will help us to understand field metabolic rates. I will present the examples in traditional taxonomic order: first molluscs, then an insect, a fish, birds, and mammals.

7.2 Mussels

Many invertebrate animals are sedentary. Their lives may seem uneventful, in comparison with those of animals that move around, but we should

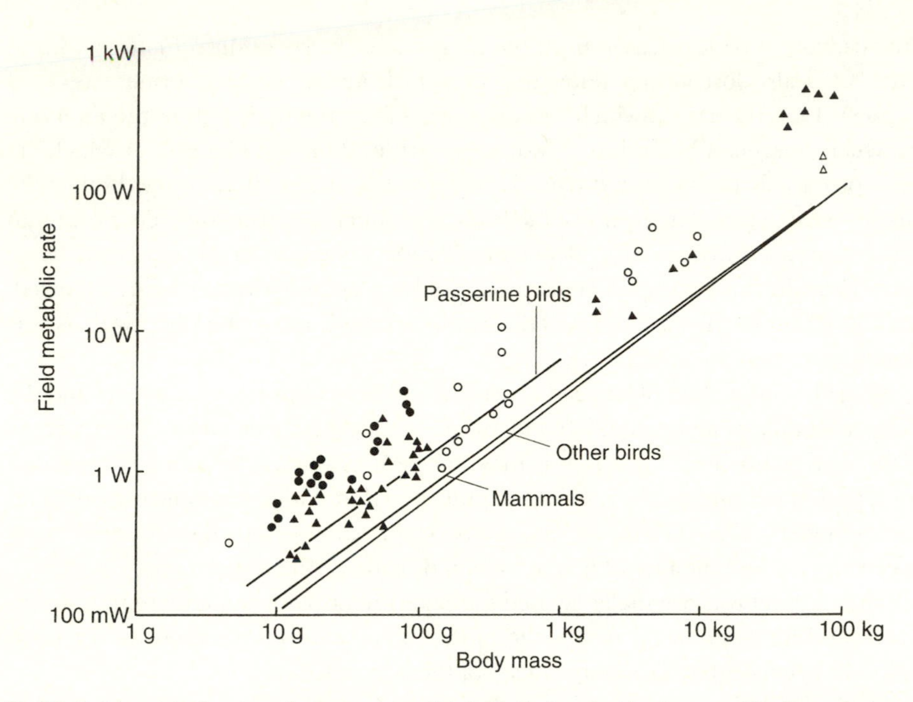

Fig.7.1 Field metabolic rates of mammals and birds plotted against body mass. The lines show predicted minimal metabolic rates (Table 2.2). Filled triangles, eutherian mammals excluding humans; open triangles, humans; filled circles, passerine birds; open circles, non-passerine birds. (Data from K. A. Nagy (1987) *Ecological Monographs* **57**, 111–28 and (for humans) J. G. V. A. Durnin (1985) *Proceedings of the Nutrition Society* **44**, 273–82.)

nevertheless ask what they do and how they use their energy, in their natural habitats.

Mussels (*Mytilus edulis*) live between the tidemarks on rocky Atlantic and Mediterranean shores and attached to man-made structures such as piers. They do not bury themselves like many other bivalves, but remain exposed to view, anchored to the rock or pier by their byssus threads. They are farmed because they are good to eat, which has been a strong incentive for research.

Mussel metabolic rates have not been measured while they remained in their natural habitat, but there are a few measurements that seem to give reliable field metabolic rates. Freshly collected mussels have been kept on a research ship, anchored over the natural population. Water pumped from just above the natural population flowed through their tanks, so they had the same water (the same temperature, salinity, and food content) as the natural population. Metabolic rates and rates of feeding, growth, and gamete production were measured.

One of the sites was a gravel bank in an English estuary. Metabolic rates were highest in summer, about 2.5 times the lowest (midwinter) rates. This was apparently due to a combination of two effects. First, the water was warmer in summer (16–20°C, compared with 7–10°C in winter). Remember the general rule that the metabolic rates of ectotherms roughly double for a 10 K temperature rise (Section 6.1). Second, more food was available in summer, making growth possible. In winter, there was too little food to support the minimal metabolic rate and the mussels survived by using some of their reserves, losing body mass. In summer, food was often plentiful and the mussels were able to grow fast. Remember that growth costs energy (Section 5.2). In June, July, and August, when their metabolic rates averaged $10\,J\,h^{-1}$, 1 g (dry mass) mussels were producing new tissue and gametes at an average rate of $17\,J\,h^{-1}$. The metabolic energy cost of growth is about 40 per cent of the energy incorporated, so this rate of production requires metabolism at a rate of $7\,J\,h^{-1}$. More than half the metabolic rate in summer must have been required to support production. Gamete production and growth of other tissues each accounted for about half the annual production of these mature mussels.

These particular mussels were low on the shore, where they were exposed to the air only at very low tides. Observations of the animals in the field showed that, while they were submerged, they kept their shells open and were presumably feeding for at least 90 per cent of the time. When exposed by the tide, mussels clam up and their metabolic rate falls. Low tide has been simulated in laboratory experiments as described in Section 2.7—these experiments indicate that a mussel exposed for 5–6 h will save around 80 per cent of the metabolic energy it would otherwise have used if it had remained submerged and open. In some of these experiments it has been shown that as the metabolic rate falls in a closed mussel, the heart slows down, typically to one third of its original frequency. This has been done by means of electrodes, inserted through fine holes in the shell on either side of the heart and fixed in place by wax. These detect the electrocardiogram, the pattern of changes of electrical potential that accompany the heartbeat.

Mussels fitted with electrodes have been returned to the field. As in the laboratory experiments, heartbeat frequency was high while they were submerged and low when they were exposed at low tide. Metabolic rates are presumably low in exposed mussels in the field, as in the laboratory.

7.3 ***Nautilus***

Nautilus is a cephalopod, a relative of the squids, but has a spiral shell with gas-filled chambers that give the animal the same density as the sea water it lives in. It inhabits coastal waters in the Pacific Ocean, at depths down to 500 metres. Like squid (Section 4.3) it swims by jet propulsion, varying its speed by using stronger or weaker jets.

Dr Ron O'Dor and his colleagues realized that jet pressure could be used as an indicator of metabolic rate. In preliminary experiments in an aquarium, they glued instruments to the shells of *Nautilus*: a pressure transducer with a cannula leading into the mantle cavity and a transmitter that emitted pulses of ultrasound that signalled the pressure in coded form. The instruments were enclosed in a streamlined cowling of plastic foam (Fig. 7.2). The density of instruments plus cowling matched the density of sea water, so that though the assembly was quite bulky it did not affect the animal's buoyancy. The animal swam against the current in a water tunnel, at various speeds, and its oxygen consumption was measured. An example of a pressure record is shown in Fig. 7.3. In this example, the animal was initially swimming at 0.27 m s^{-1}. The pressure peaks were close to 2 kPa but the **mean** pressure, taking account of the troughs as well as the peaks, was only 0.51 kPa. The investigators showed that mean pressure, calculated in this way, correlated well with swimming speed and metabolic rate.

Nautilus with the equipment attached were returned to their natural habitat. They were followed in a boat, with a receiver to record the ultrasonic signals. Changes in absolute pressure were obtained from the transducer, as well as pressure differences between the mantle cavity and the water outside, so a record of the animal's depth was obtained. Its position was also determined, using the military Global Positioning System. The records showed that the animals were a little more active at night than by day. They tended to swim upward at dusk and down again at dawn, through vertical ranges of 50 metres or more. Upward and downward swimming each used about the same metabolic power as horizontal swimming at the same speed; this is what one should expect, for an animal with the same density as the water. On one occasion an animal travelled a horizontal distance of 1 km in 24 h. The mean metabolic rate over 24 h, estimated from the

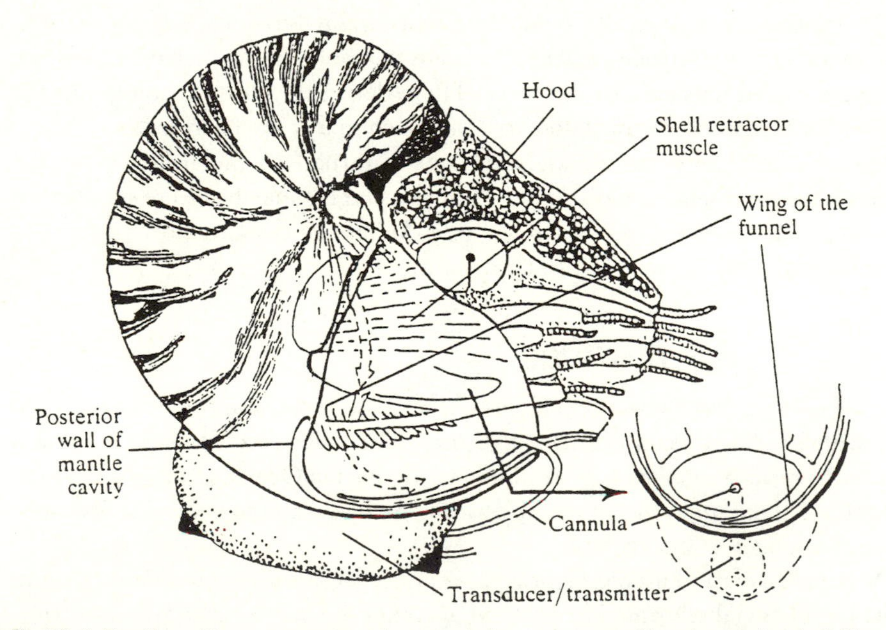

Fig. 7.2 A *Nautilus* with a pressure transducer and ultrasonic transmitter glued to its shell. The cannula is placed in the mantle cavity. The transducer senses the difference in pressure between the mantle cavity and the water outside. (From R. K. O'Dor, J. Wells, and M. J. Wells (1990) *Journal of Experimental Biology* **154**, 383–96.)

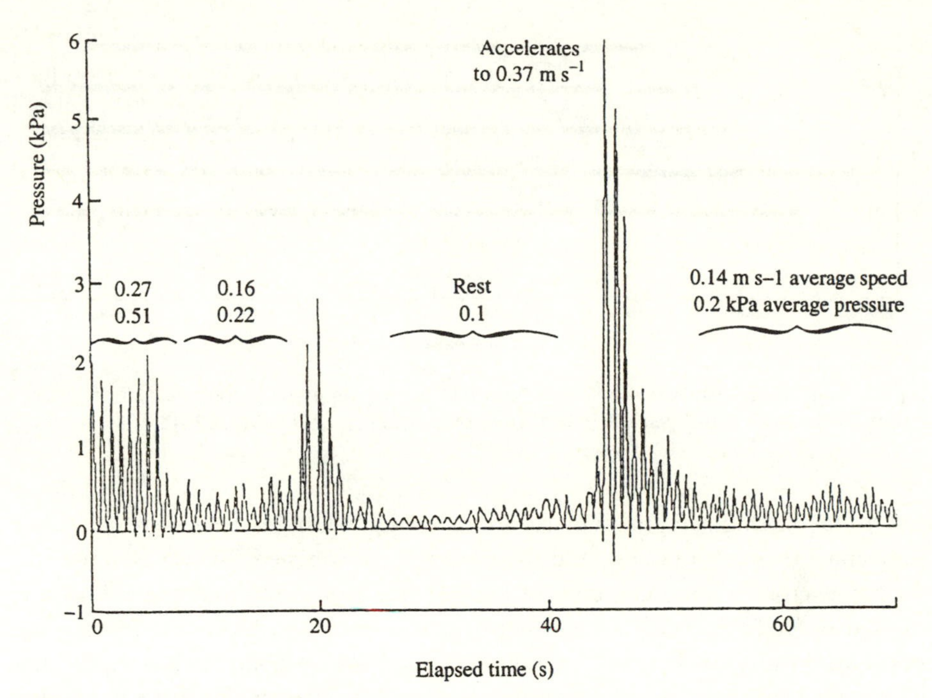

Fig. 7.3 A record of pressure in the mantle cavity of a captive *Nautilus* swimming spontaneously, obtained from the equipment shown in Fig. 7.2. (From R. K. O'Dor, J. Wells, and M. J. Wells (1990) *Journal of Experimental Biology* **154**, 383–96.)

record of mantle pressures, was 0.17 W kg^{-1}. For comparison, the minimal (resting) and maximal (swimming fast) rates measured in the laboratory were about 0.06 and 0.56 W kg^{-1}. Thus the field metabolic rate was about three times the minimal rate.

7.4 **Bumble-bees**

Bees are proverbially busy. A study of a bumble-bee species (*Bombus vosnesenskii*) in California showed that the large workers (the ones that forage for nectar and pollen) work without a break throughout the hours of daylight in summer. Each colony has a nest in a burrow, with a single entrance. Members of a colony were captured and fitted with coloured and numbered tags, big enough for an observer to read but too small to cause the bee inconvenience. Observers watched the burrow entrance for a whole day, from 5 a.m. to 9 p.m., and clocked the marked bees in and out. Figure 7.4 shows the exits and entrances of five workers. Their foraging trips were 30–90 minutes long, with only 5 minutes in the burrow between trips. Even that 5 minutes cannot have been rest—subsequent excavation showed that the tunnel was long, with the nest (where the bees deposited their loads) 2 metres from the entrance.

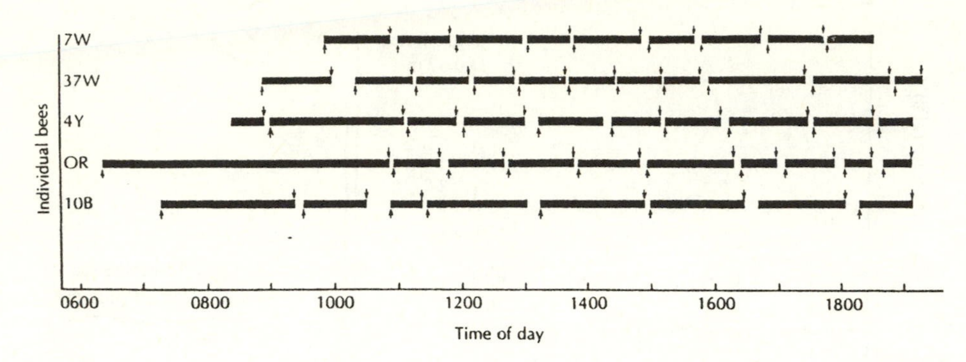

Fig. 7.4 Times at which individual bumble bees left (↑) and re-entered (↓) their nest, during 1 day. (From T. Allen *et al.* (1978) *Journal of the Kansas Entomological Society* **51**, 329–42.)

On the day following the one recorded in Fig. 7.4, a few bees were caught as they entered or left the nest, and their loads were weighed. Nectar is carried in the honeycrop, a large chamber of the gut that lies in the abdomen. Pollen is carried packed between rows of long spines on the hind legs. Bees leaving the nest carried tiny sugar loads, averaging 2 mg, and no pollen. Those returning had sugar loads averaging 27 mg, and 50 per cent of them also carried a pollen load averaging 21 mg. The average total load was about 60 per cent of unloaded body mass.

The bees foraged in a very systematic way. Individuals specialized on different flower species. The species they were visiting had different colours of pollen, but each bee consistently brought the same colour. Further observations showed that individual bumble-bees often visit the same clumps of flowers in the same order, day after day.

The doubly labelled water technique has recently been adapted for use with bumble-bees; we may hope soon to have good measurements of field metabolic rates. Meanwhile we can try some estimates. We saw in Section 4.2 that flying bumble-bees have metabolic rates around 350 W kg^{-1} or 1200 J g^{-1} h^{-1}, at all speeds of flight. Other experiments have shown that when a load is carried, metabolic rate is increased roughly in proportion to the total loaded mass. Thus a bee returning to the nest with a typical load may be expected to have a metabolic rate of around 1800 J (g unloaded mass)$^{-1}$ h^{-1}. Metabolic rates of inactive bees that have allowed their bodies to cool to the temperature of their surroundings vary considerably, depending on what that temperature is, but are generally around 50 J g^{-1} h^{-1}. Bees perched on plants while foraging may use no more energy than this on warm days, but on cold days they would have to shiver to avoid cooling to a temperature at which they could not fly. We will assume a warm day.

To work out an energy budget we need to know how much time is spent on each activity, as well as the metabolic rate for each. Figure 7.4 shows that on a particular day at a particular place, bumble-bee workers foraged for about 10 h

Table 7.1 A possible energy budget for a foraging bumble-bee worker

Activity	Duration (h)	Metabolic rate ($J\ g^{-1}\ h^{-1}$)	Energy cost ($J\ g^{-1}\ day^{-1}$)
Unloaded flight	2.5	1200	3000
Loaded flight	2.5	1800	4500
On flowers	5	50	250
Resting	14	50	700
Total	24		8450

and spent 14 h, presumably resting, in the burrow. How much of the foraging time is spent flying depends on the species of flower (some can be processed faster than others) and on how far the flowers are from the nest. Data collected by Dr Bernd Heinrich (who is responsible for most of the information in this section) suggest that an average of 50 per cent of foraging time is spent in flight, so we will assume our bees spend 5 h on flowers and 5 h actually flying. During each foraging trip the load will increase gradually, but we can estimate energy costs by assuming 2.5 h unloaded and 2.5 h fully loaded flight.

These assumptions give us Table 7.1 as a first attempt at a 24 h energy budget for a foraging worker. It indicates that during a warm summer day, a foraging bumble-bee might use 8450 joules per gram body mass. If it had rested all day it would have used only an estimated 1200 joules per gram body mass. Thus the field metabolic rate is estimated to be seven times the minimum rate. This is an exceedingly rough calculation, but the conclusion matches the general impression, that bumble-bees lead very active lives. They certainly seem more active than mussels and *Nautilus*, whose field metabolic rates seem to be much smaller multiples of their minimal rates.

7.5 **Trout**

We have already seen, in Section 4.3, how the metabolic rates of fish swimming at different speeds have been measured. Now we ask, how much of their time do fish spend swimming at different speeds, using energy at different rates? Some pioneering experiments were done on trout (*Salmo trutta*) in a lake in Scotland.

This research depended on the assumption that heartbeat frequency is a reliable indicator of metabolic rate. The faster an animal is using oxygen, the more rapidly must the blood carry oxygen to the tissues, and the faster the heart can be expected to beat. It was known from previous experiments with trout that there was a clear relationship between metabolic rate and heartbeat frequency; electrodes fitted to trout detected the electrical signals which accompany the heartbeat, while the fish swam in a water tunnel and their rates of oxygen consumption were measured.

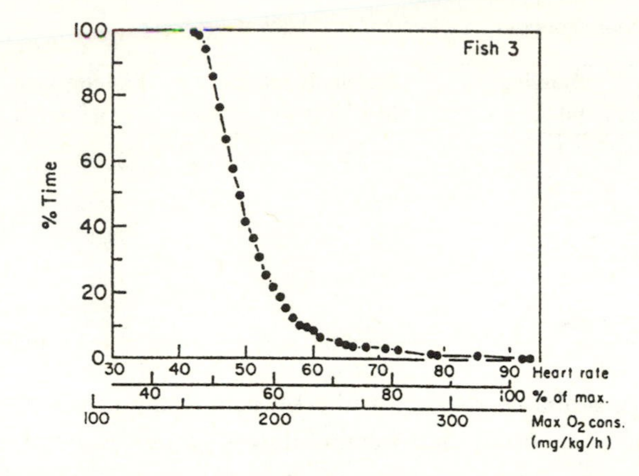

Fig. 7.5 The distribution of heartbeat frequencies (min^{-1}) observed for a trout in a lake in Scotland, during a period of 3 days. Note that the highest frequencies occur for only a tiny fraction of the time. A scale at the bottom shows estimated rates of oxygen consumption. The minimum metabolic rate at the temperature of the lake (15°C) would be about 120 mg O_2 kg^{-1} h^{-1} (0.45 W kg^{-1}). (From I. G. Priede and A. H. Young (1977) *Journal of Fish Biology* **10**, 299–318.)

The observations on trout in the lake used an ultrasonic tag 39 mm long (10 per cent of the length of the fish). A fish was caught and anaesthetized. A tag was wired to the dorsal fin and connected to electrodes near the heart. It emitted an ultrasonic bleep at every heartbeat. When the fish had recovered, it was returned to the lake. The bleeps were received by hydrophones, placed at three points in the lake, which could be used to locate the fish as well as to record its heartbeats. Observations continued for as long as the tag's battery lasted, for up to 7 days. The fish patrolled restricted areas of the lake, up to about 150 m diameter. Their heartbeat frequencies were consistently lower at night than during the day, indicating a lower level of activity. Figure 7.5 shows that the maximum heartbeat frequency (corresponding to the highest attainable metabolic rate) was very rarely used. The mean metabolic rate was estimated to be 1.6 times the minimal rate.

The main conclusion from this study probably applies widely—animals generally exert maximum effort for only a tiny proportion of the time.

7.6 House martins

Professor David Bryant and his colleagues used the doubly labelled water technique to measure the metabolic rates of house martins (*Delichon urbica*), nesting in Scotland. These are small birds of about 20 g body mass that feed on insects, which they catch in flight. Natural nests are built of mud, under the eaves of houses or on cliff faces, but the birds that were studied were using specially designed nest boxes. These had trap doors that could be used to capture the birds as required, to inject doubly labelled water or to take a blood sample. Metabolic rates were measured over periods of 1 or 2 days. While that was

Table 7.2 Metabolic rates of house martins at three stages of the breeding season

	Non-breeding	Incubating	Feeding nestlings
Metabolic rate/minimal rate	2.7	2.8	3.6
Hours per day off nest	11.3	6.4	14.1
Flapping time as % flying time	33%	46%	40%

Data are from D. M. Bryant and K. R. Westerterp (1980) *Ardea* **68**, 91–102.

happening, the nest box was watched throughout the daylight hours and the times of all arrivals and departures were noted. Also, for part of the time the birds were watched feeding; dye marks had been put on the white parts of their plumage making them individually identifiable. They were never seen perching, except when they landed at the nest. They flew partly by flapping and partly by gliding, and records were made of the fraction of flying time spent flapping.

Data are shown in Table 7.2. The 'non-breeding' column shows metabolic rates measured before any eggs were laid, or after the nestlings had flown. The other columns show rates during incubation and while young birds were being fed in the nest. The parents share the tasks of incubation and feeding the young. No data are shown for the period of egg laying because females injected in this period interrupted laying, presumably as a result of the disturbance. In other periods, differences in metabolic rate between the sexes were small, so data for males and females have been lumped together. Flying insects were plentiful (as shown by catches in a suction trap) throughout the breeding season.

During the non-reproductive periods, the birds had plenty of time to get their food. There was no need to stay at the nest to incubate and they had only themselves to feed. They spent almost all the daylight hours off the nest, airborne all the time, but 67 per cent of their flying time was gliding, with only 33 per cent of more energy-demanding flapping flight. During incubation, the birds still had only themselves to feed, but they took turns to incubate the eggs, so each had only half the daylight hours available for feeding. Having to feed faster, their flight was less leisurely, with 46 per cent of flapping. More strenuous flying for less of the day resulted in a metabolic rate only a little higher than in the non-breeding periods. Finally, while there were young in the nest the parents again had all day to feed, but they had to get food for the young as well as themselves. Each made about 100 visits to the nest during the day, bringing mouthfuls of insects for the young. The quantity of insects in each mouthful was measured in separate experiments in which nestlings were temporarily fitted with collars that prevented swallowing, enabling the experimenters to recover the mouthful. This led to a calculation that the quantity of food collected by each parent was enough to support a total metabolic rate of 6.6 times the minimal adult metabolic rate, 3.6 times minimal for the parent itself, and the balance for the nestlings. The food collected and the flying time were 2.3 and 2.2 times higher than during incubation, respectively, so the rate of catching insects was almost unchanged. The increased feeding effort explains the increased metabolic rate. Though the birds flew for 2.2 times as long, their flight may not have been quite as strenuous as during incubation; they flapped for only 40 per cent of the time, possibly because they could glide for much of the time when commuting between the nest and the feeding area which was sometimes more than a kilometre away.

7.7 **Albatrosses**

Albatrosses soar over the Pacific and Southern Oceans. The black-browed albatross *Diomedea melanophrys* is one of the smaller species, with a mass of about 4 kg, but even a small albatross is a large bird. The research I am going to describe was done on South Georgia, an island where black-browed albatrosses nest far south in the Atlantic. The aim was to discover how the birds use their time and energy.

In preliminary experiments, black-browed albatrosses had been trained to walk on a treadmill while their rates of oxygen consumption and heartbeat frequencies were measured, the latter by means of electrodes which recorded the electrocardiogram. The treadmill was run at different speeds and a graph of heartbeat frequency against metabolic rate obtained. This graph was used to estimate metabolic rates from heartbeat frequencies recorded in the field; it was assumed that a given heart rate would correspond to the same metabolic rate, whatever the bird was doing.

Albatrosses were caught at their nests during the Antarctic summer, between November and February. They were anaesthetized and fitted with recording equipment, then returned to the nest. During the 2 hours needed for this, the egg or chick was cared for. Twenty-five birds were fitted with a data logger (a tiny computer) which was implanted surgically in the abdominal cavity. Connected to the data logger were electrocardiograph electrodes and a thermistor that sensed the temperature in the abdominal cavity. While in the bird, the data logger recorded the heartbeat frequency every 30 seconds and the temperature every minute. Thirteen to 33 days later, the bird was recaptured at its nest, the data logger was retrieved, and the recorded data were downloaded into a portable computer.

Three other devices were fitted to a few birds. Five birds had salt-water switches attached to rings on their legs. These recorded the times when the bird was on the sea surface, by sensing the resistance between two electrodes. Two birds were fitted with tiny radio transmitters that made it easy to detect their presence on the island, and with separate transmitters that enabled a satellite to locate them, even when they were hundreds of kilometres from the island.

Albatrosses nest on the ground, laying just one egg. The parents take turns at the nest, one incubating the egg or brooding the young chick while the other forages for food. Later, when the chick is older, both parents go foraging, returning only briefly to feed the chick. Figure 7.6 is a record from the brooding

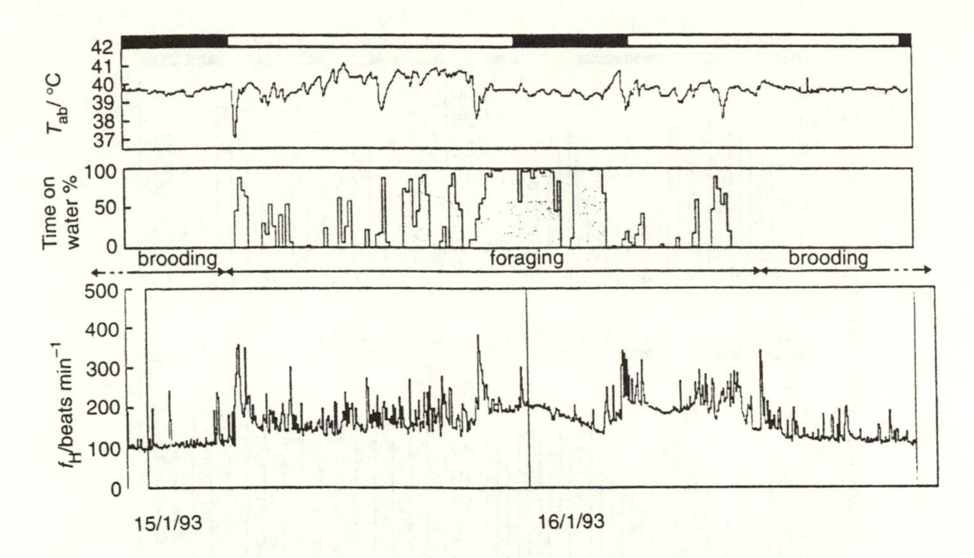

Fig. 7.6 A record from an instrumented black-browed albatross showing abdominal temperature, time on the sea surface, and heartbeat frequency during 2 days. Black bars at the top show the hours of darkness. (From R. M. Bevan *et al.* (1995) *Proceedings of the Royal Society* B **350**, 119–31.)

period. The instrumented bird left the nest at dawn on 15 January and returned at noon on the following day. The bar at the top of the figure indicates daylight and darkness. The 'time on water' record shows that the bird spent quite a lot of its foraging time on the water, including most of the night. Albatrosses often catch food such as squid and krill while floating on the sea surface; it is at night that these animals come to the surface. On average, foraging birds spent 70 per cent of their time in the air and 30 per cent on the water. The temperature record shows a more or less constant temperature while the bird was on the nest and brief pulses of reduced temperature while it was foraging. These occur when the bird was on the water and presumably show when it swallowed cold food. The heartbeat frequency shown in the bottom record is relatively low while the bird was on the nest, but more variable and generally higher when it was foraging.

Figure 7.7 shows metabolic rates estimated from the heartbeat frequency. In this figure, the bars at the tops of the graphs show when the birds were at their nests. During incubation, foraging trips were relatively long; the average for all the instrumented birds was 9 days. While the chick was being brooded, trips were shorter, averaging only 2.5 days. At the nest, heartbeat frequency was relatively low, indicating metabolic rates averaging 2.3 W kg^{-1} (7 cm^3 O_2 min^{-1} kg^{-1}). During foraging, the average metabolic rate was about 5 W kg^{-1}, with no significant difference between times when the bird was airborne and times when it was on the water. Soaring albatrosses hardly ever flap their wings, so their energy consumption is low (see Section 4.2). On the water surface they are generally

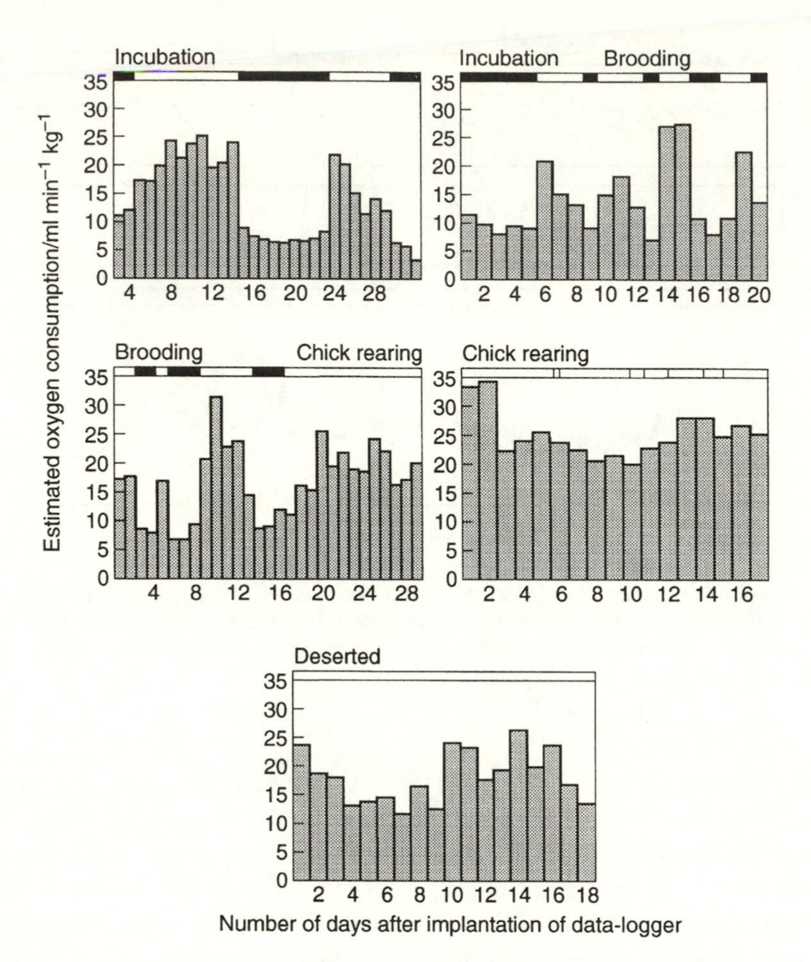

Fig. 7.7 Rates of oxygen consumption estimated from records of heartbeat frequency of individual black-browed albatrosses, plotted against the number of days since the data logger was implanted. Black bars along the tops of the graphs show when the bird was at its nest. The bottom graph refers to a bird that deserted its nest. (From R. M. Bevan *et al.* (1995) *Proceedings of the Royal Society* B **350**, 119–31.)

fishing, so are quite active. It is coincidental but not too surprising that the average rates of use of energy are about the same for soaring and fishing.

The visits to the nest at the chick-rearing stage averaged only 11 minutes and would often have been missed if observation had been visual. They were registered by the recording receiver, for the birds that carried radio transmitters.

Straight line soaring speeds (measured in a different investigation) averaged 13 m s^{-1} (47 km h^{-1}). Albatrosses generally soar along zig-zag paths, but nevertheless travel long distances in a foraging trip of several days.

7.8 Seals

Another study of the everyday activity of marine animals was of grey seals (*Halichoerus grypus*) off the west coast of Scotland. These are animals of around 150 kg body mass, so making equipment small enough for them to carry was less of a problem than for the animals discussed earlier in this chapter. Seals caught in nets were immobilized and fitted with several devices, which were glued to their skin or fur:

(1) a pair of electrodes attached near the heart, which sensed the electrocardiogram;
(2) a pressure transducer whose output indicated the depth at which the seal was swimming; and
(3) a paddlewheel flow meter, which indicated the rate of flow of water past the seal, indicating the speed at which it was swimming. This could not detect speeds below about 0.4 m s^{-1}.

The data from these instruments were transmitted continuously as a coded ultrasonic signal, which was received by a hydrophone on a yacht at a distance of a few hundred metres. In addition, a radio transmitter glued to the seal's head was useful for locating it from greater distances, and gave a clear indication of surfacing. Unlike the ultrasonic signals, which travelled through the water, the radio signal could travel only through air and was received only when the seal's head was above water.

Figure 7.8 is a typical record of 5 hours' activity. Heartbeat frequencies are high (around 130 min^{-1}) when the depth record shows that the seal was at the surface, and low (around 30 min^{-1}) when it was submerged. The swimming speed was too low to be registered whenever the seal was at the surface. Visual observations confirmed that seals at the surface did not swim, except occasionally when inspecting beaches. Visits to the surface were brief, averaging about 50 seconds.

The record shows two distinct kinds of dive. Most dives were short (less than 10 min), with the seal swimming continuously at 1–2 m s^{-1}. Occasional dives were longer, up to 26 min in another record. In these, the animal dived immediately to or near the bottom and remained there motionless or swimming very slowly. The short dives were the animals' normal mode of travelling. Diving so deep while travelling may seem wasteful of energy, but the depths of around 50 metres are small compared to the distances travelled—a seal swimming at 1.5 m s^{-1} during a typical 7 min dive would travel 630 m. Remember that for a neutrally buoyant swimmer, swimming upwards costs no more energy than swimming horizontally. The seals may have been inspecting the bottom as they travelled. During the longer dives the seals were presumably foraging, either slowly searching the bottom for food or lying in ambush for suitable prey. Grey seals feed largely on fish such as cod, salmon, and herring.

The longer a dive, the longer you might expect the seal to spend breathing on

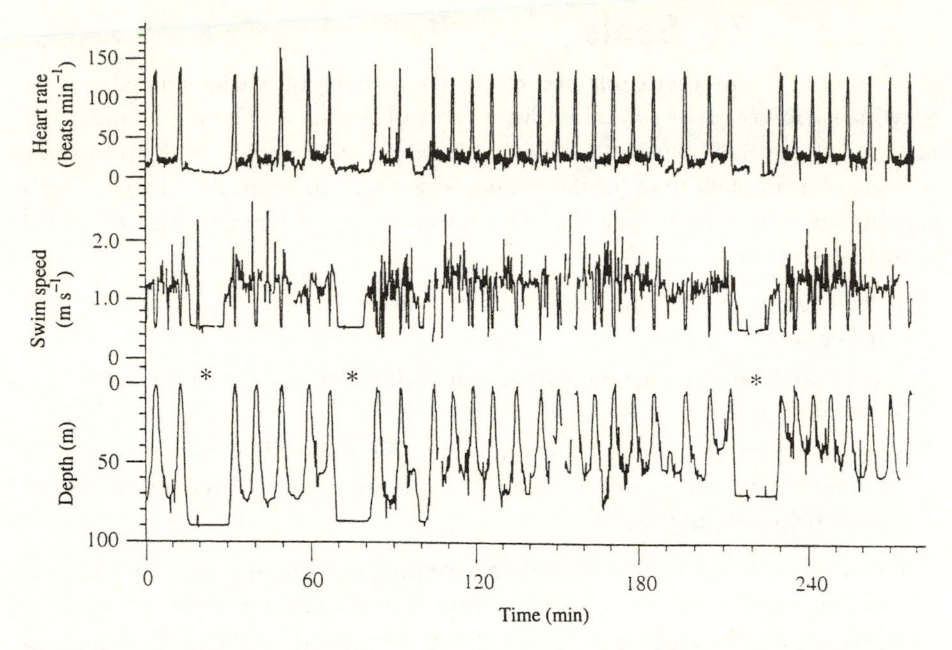

Fig. 7.8 Records of heartbeat frequency, swimming speed, and depth for a grey seal swimming between haul-out sites. The time scale is calibrated in minutes. *, pauses when swimming stopped. (From D. Thompson and M. A. Fedak (1993) *Journal of Experimental Biology* **174**,

its next visit to the surface. Surface visits after very short dives are indeed short, but after dives of over 7 min there is no further increase in the time spent at the surface. This suggests that even in the longest dives the animal does not accumulate an oxygen debt. Stopping swimming during long dives may be an oxygen-saving tactic. Another suggestive observation is that, in the longest dives, heartbeat frequency falls very low indeed. However long the dive, there are never more than about 220 heartbeats in it.

No estimates of energy costs were made in this study, but the animals' diving behaviour seems to have been constrained by the need to conserve energy in long dives, because oxygen supplies are limited.

Grey seals do not spend all their time swimming and diving as described in this section. A great deal of their time is spent hauled out on beaches, or resting in the water near haul-out sites.

7.9 Mountain farmers

Many studies have been made of how people with different life-styles spend their time and use their energy. I have chosen as an example a group of farmers whose life gives the impression of being unusually demanding. They live in Nepal, in the foothills of the Himalayas, in a village 1870 metres above sea level. They take advantage of their situation on a mountainside to cultivate crops at different levels, from 1350 to 3800 m, which ripen at different times and can be harvested in succession. They grow rice, other cereals, and tubers, keep cattle and other animals, and collect firewood and fodder in the forest. This makes them largely self-sufficient, needing few supplies from outside the village. The work done by the women has been studied in most detail.

On a few days, spaced out through the year, each woman was followed throughout the daylight hours by a trained observer who noted down everything she did, with times. In order to disturb the women's normal routine as little as possible, the observers were recruited from among the villagers and were instructed to watch from a distance. Table 7.3 shows how women who were neither pregnant nor lactating spent their time. January to March is late winter, with relatively little agricultural work. July to September is the monsoon season when the workload is heaviest. Travel times are long because the cultivated terraces extend considerable distances up and down the mountain. The women were assumed to be resting or asleep during the hours of darkness, when they were not watched.

The rates of oxygen consumption of women were measured while they performed representative tasks. For some of these measurements a Douglas bag was used; this is a bag worn as a backpack which collects expired air for analysis. For others, a different type of respirometer was used. The examples of energy costs shown in the table are hoeing fields; chopping wood with a kukri (a large knife); walking up or down hill with moderately heavy (10–39 kg) loads, carried using a head yoke; pounding grain; and resting. Though many of these tasks look backbreaking, the higher metabolic rates fall in the range generally classified as

Table 7.3 Time and energy allocation by women in a village in Nepal

	Activity	**Time (minutes per day)**		**Energy cost/basal metabolic rate**
		January–March	**July–September**	
Outdoors	Agriculture	81	306	Hoeing 5.1
	Husbandry	61	41	
	Forest work	19	0	Chopping 4.0
	Travel	83	58	Uphill 5.6
				Downhill 3.6
	Rest	81	90	1.3
Indoors	Domestic work	161	115	Pounding 4.8
	Family care	49	40	
	Walking about	15	12	
	Rest	136	102	1.3
	Not watched	754	676	
Mean metabolic rate/basal		1.89	2.01	

Data are from C. Panter-Brick (1993) *American Journal of Clinical Nutrition* **57**, 620–8.

moderate rather than heavy work. Rests were taken frequently, for example for about 10 minutes in each hour of walking. Also, women carrying heavy loads uphill slowed down to such an extent that their metabolic rates were less than when carrying lighter loads.

These and other measurements of metabolic rates were used with the diaries of activities, to estimate the mean 24 h metabolic rates, given at the bottom of the table. This method seems to work reasonably well; in a similar study in the Gambia, metabolic rates estimated by timing activities agreed well with measurements by the doubly labelled water method. The metabolic rates shown at the bottom of the table are similar to the estimate of twice the basal rate shown for a labourer in Fig. 7.1.

Pregnant and lactating women spent less time on outdoor work in winter than women who were neither pregnant nor lactating. In the monsoon season, however, all women worked approximately the same hours regardless of their condition.

7.10 **Limits to sustained metabolism**

We started this chapter with the doubly labelled water technique and its use to measure the metabolic rates of animals in their natural habitats. Various mammals and birds that have been investigated in this way have field metabolic rates (averaged over 24 h) of around three times their minimal metabolic rates. Then we saw how various animals use energy in their everyday lives. Mussels do not move around; any metabolism above the minimal rate is used for growth. For our other examples, locomotion is a large user of metabolic energy. Field metabolic rates ranged from 1.6 times the minimal rate for trout, which swam fast only rarely; to 7 times the minimal rate for bumble-bee workers, which had very high metabolic rates during their daily 5 h of flight. Metabolic rates of house-martins are considerably higher while caring for nestlings than outside the reproductive season. Albatrosses spend a very large proportion of their time flying, but fly very economically by soaring. Seals seem to be constrained by the need to conserve oxygen while diving; they may swim fast during short dives but swim slowly during long ones. For women farming in Nepal, strenuous activities such as hoeing and climbing mountainsides require 5–6 times the minimal metabolic rate, but the mean metabolic rate over 24 h is only twice the minimal rate.

In a wider survey, Drs Kimberly Hammond and Jared Diamond collected metabolic rates for wild mammals, birds, and reptiles during periods when the energy demands on them were high. For example, they found data for rodents during very cold weather, or during lactation, and for birds rearing chicks. Among the 50 examples they collected, none had metabolic rates, averaged over 24 h, greater than seven times the minimal metabolic rate. Five examples exceeded six times the minimal rate, but none reached seven. Is there a fixed limit to the metabolic rates that animals can sustain without depleting their energy reserves, and what sets the limit?

In an attempt to answer this question, Dr Hammond performed experiments with laboratory mice. She provided unlimited food and elicited high metabolic rates in several different ways. She made some mice run on treadmills until they were exhausted. They ran for 6 h each day and had metabolic rates (averaged over 24 h) of 3.6 times the minimal rate. She gave lactating mice additional pups to suckle and so drove their metabolic rates up to 6.5 times the minimal rate. Other experimenters kept mice at the lowest temperature they could withstand, –15°C, and measured metabolism at 4.8 times the minimal rate. In all these cases there seemed to be limits, but the limits were different in different conditions. This implies that the limiting processes were different—the results could not be explained by any universal limit to the rate at which the body could handle energy. Metabolism during lactation may perhaps be limited by the rate at which the gut can process food, but exercise and heat production have limits specific to them.

The circumstances in which the highest metabolic rates can be obtained are different for different animals. For laboratory mice it is lactation, but deer mice that attained only 4.6 times the minimal rate in lactation achieved 6.1 times the minimal rate in cold conditions. And the highest recorded sustained metabolic rates for humans are for cyclists on the Tour de France, who metabolized at 4.3 times the minimal rate during the 22 day race.

These data show that field metabolic rates are generally well below maximum possible sustained metabolic rates. Why do animals not exert themselves more? The answer may well be that excessive work is bad for health. In experiments in which bees foraging from a hive were made to work harder, by glueing lead weights to their thoraces, loaded bees lived less long. Unloaded bees and bees with very light loads lived a mean of 11 days after being caught, marked, and loaded; but bees with loads of 35 per cent or more of body weight lived only 8 days. In another set of experiments, kestrels (*Falco tinnunculus*) nesting in the wild had nestlings added to or removed from their broods. The mean original clutch size was 5.4; two nestlings were added to or removed from each brood. Added nestlings were similar in age to the original brood and were accepted by the parents. The parents were ringed, so could be identified if, eventually, they were found dead. Parents with increased broods had to work harder to rear them than those with reduced broods; measurements using the doubly labelled water method showed that they used 22 per cent more energy in doing so. Parents with increased broods were much more likely to die before the next breeding season than those with reduced broods; the mortality rates were 60 per cent for those with increased broods and 22 per cent for those with reduced broods.

Epilogue

There is a temptation in writing books like this to try to make every story seem complete and well understood. But it seems right at the end of this particular book to emphasize that our knowledge of the role of energy in animal life is very imperfect. We cannot explain quantitatively the minimal metabolic rate; it seems clear that protein turnover and ion pumping account for a good deal of it, but what determines the rates at which proteins have to be turned over, and why are cell membranes not made less leaky so that ions need not be pumped so fast? Many attempts have been made to explain the very general rule, that metabolic rates of related animals tend to be proportional to (body mass)$^{0.75}$, but even the latest and best explanation does not seem wholly satisfactory. We do not understand how energy is used in muscle sufficiently well to make precise, confident predictions of the metabolic costs of the movements that are powered by muscles. The theory of life history strategies is an elegant mathematical structure, with remarkably little foundation in solid biological data for particular species. It is not clear why the high (wasteful?) metabolic rates of birds and mammals are advantageous; in many respects, lizards heated by the sun perform as well. And there are very few animals for which we have clear, quantitative knowledge of how they use energy in their everyday lives. Even the examples in Chapter 7 (the best examples I could find) are very incomplete.

Bibliography

Chapter 1

Archibold, O. W. (1995). *Ecology of world vegetation.* Chapman & Hall, London.

Grahame, J. (1987). *Plankton and Fisheries.* Arnold, London.

Hay, R. K. M. and Walker, A. J. (1989). *An introduction to the physiology of crop yield.* Longman, London.

Horn, H. S. (1971). *The adaptive geometry of trees.* Princeton University Press, Princeton.

Lewis, J. B. (1977). Processes of organic production on coral reefs. *Biological Reviews* **52**, 305–47.

Monteith, J. L. and Unsworth, M. H. (1990). *Principles of environmental physics* 2nd edn. Arnold, London.

Raven, P. H., Evert, R. F., and Eichhorn, S. E. (1992). *Biology of plants*, 5th edn. Worth, New York.

Scott, K. M. and Fisher, C. R. (1995). Physiological ecology of sulfide metabolism in hydrothermal vent and cold seep vesicomyid clams and vestimentiferan tube worms. *American Zoologist* **35**, 102–11.

Chapter 2

Blaxter, K. (1989). *Energy metabolism in animals and man.* Cambridge University Press.

Elgar, M. A. and Harvey, P. H. (1987). Basal metabolic rates in mammals: allometry, phylogeny and ecology. *Functional Ecology* **1**, 25–36.

Else, P. L. and Hulbert, A. J. (1987). Evolution of mammalian endothermic metabolism: 'leaky' membranes as a source of heat. *American Journal of Physiology* **253**, R1–7.

Martin, A. W. and Fuhrman, F. A. (1955). The relationship between summated tissue respiration and metabolic rate in the mouse and dog. *Physiological Zoology* **28**, 18–34.

Peters, R. H. (1983). *The ecological implications of body size.* Cambridge University Press.

Torres, J. J., Belman, B. W., and Childress, J. J. (1979). Oxygen consumption rates of midwater fishes as a function of depth of occurrence. *Deep-Sea Research* **26A**, 185–97.

West, G. B., Brown, J. H., and Enquist, B. J. (1997). A general model for the origin of allometric scaling laws in biology. *Science* **276**, 122–6.

Wieser, W. and Gnaiger, E. (eds) (1989). *Energy transformations in cells and organisms.* Thieme, Stuttgart. (See articles by Chinet on ion pumping in muscle, by Widdows on mussels, and by Waterlow and Millward on protein turnover.)

Chapter 3

Alexander, R. McN. (1996). *Optima for animals*, 2nd edn. Princeton University Press.

Alsop, D. H. and Wood, C. M. (1997). The interactive effects of feeding and exercise on oxygen consumption, swimming performance and protein usage in juvenile rainbow trout (*Onchorhynchus mykiss*). *Journal of Experimental Biology* **200**, 2337–46.

Beddington, J. R. and May, R. M. (1982). The harvesting of interacting species in a natural ecosystem. *Scientific American* **247**, 42–9.

Blaxter, K. (1989). *Energy metabolism in animals and man.* Cambridge University Press.

Chivers, D. J. and Langer, P. (eds) (1994). *The digestion system in mammals: food, form and function.* Cambridge University Press.

Colinvaux, P. (1980). *Why big fierce animals are rare.* Allen & Unwin, London.

Hughes, R. N. (ed.) (1993). *Diet selection.* Blackwell, Oxford.

Sinclair, A. R. E. and Norton-Griffiths, M. (eds) (1979). *Serengeti: dynamics of an ecosystem.* University of Chicago Press.

Stephens, D. W. and Krebs, J. R. (1986). *Foraging theory.* Princeton University Press.

Stevens, C. E. (1988). *Comparative physiology of the vertebrate digestive system.* Cambridge University Press.

Chapter 4

Alexander, R. McN. (1992). *Exploring biomechanics: animals in motion.* Scientific American Library, New York.

Alexander, R. McN. (ed.) (1992). *Mechanics of animal locomotion.* Springer, Berlin.

Azuma, A. (1992). *The biokinetics of swimming and flying.* Springer, Tokyo.

Brackenbury, J. (1992). *Insects in flight.* Blandford, London.

Brodsky, A. K. (1994). *The evolution of insect flight.* Oxford University Press.

Dial, K. P., Biewener, A. A., Tobalske, B. W., and Warrick, D. R. (1997). Mechanical power output of bird flight. *Nature* **390**, 67–70.

Ellington, C. P. and Pedley, T. J. (eds) (1995). *Biological fluid dynamics*. Symposia of the Society for Experimental Biology Vol. 49, p. 363. Company of Biologists, Cambridge.

Hedenström, A. and Alerstam, T. (1995). Optimal flight speed of birds. *Philosophical Transactions of the Royal Society* B **348**, 471–87.

Maddock, L., Bone, Q., and Rayner, J. M. V. (eds) (1994). *Mechanics and physiology of animal swimming*. Cambridge University Press.

McMahon, T. A. (1984). *Muscles, reflexes, and locomotion*. Princeton University Press, Princeton NJ.

Pennycuick, C. J. (1997). Actual and 'optimum' flight speeds: field data reassessed. *Journal of Experimental Biology* **200**, 2355–61.

Rayner, J. M. V. (1995). Flight mechanics and constraints on flight performance. *Israel Journal of Zoology* **41**, 321–42.

Suarez, R. K. (1996). Upper limits to metabolic rates. *Annual Reviews of Physiology* **58**, 583–605.

Videler, J. J. (1993). *Fish Swimming*. Chapman and Hall, London.

Weibel, E. R. and Taylor, C. R. (eds) (1981). Design of the mammalian respiratory system. *Respiration Physiology* **44**, 1–164.

Woledge, R. C., Curtin, N. A., and Homsher, E. (1985). *Energetic aspects of muscle contraction*. Academic Press, London.

Chapter 5

Alexander, R. McN. (1996). *Optima for animals*, 2nd edn. Princeton University Press.

Blaxter, K. L. (1962). *The energy metabolism of ruminants*. Hutchinson, London.

Blaxter, K. (1989). *Energy metabolism in animals and man*. Cambridge University Press.

Charnov, E. L. (1993). *Life history invariants. Some explorations of symmetry in evolutionary biology*. Oxford University Press.

Sibly, R. M. and Calow, P. (1986). *Physiological ecology of animals: an evolutionary approach*. Blackwell, Oxford.

Stearns, S. C. (1992). *The evolution of life histories*. Oxford University Press.

Chapter 6

Altringham, J. D. and Block, B. A. (1997). Why do tuna maintain elevated slow muscle temperatures? Power output of muscle isolated from endothermic and ectothermic fish. *Journal of Experimental Biology* **200**, 2617–27.

Bradshaw, S. D. (1986). *Ecophysiology of desert reptiles*. Academic Press, Sydney.

Carey, F. G. (1982). Warm fish. In *A companion to animal physiology* (ed. C. R. Taylor, K. Johanson, and L. Bolis), pp. 216–33. Cambridge University Press.

Casey, T. M. (1992). Biophysical ecology and heat exchange in insects. *American Zoologist* **32**, 225–37.

Denny, M. W. (1993). *Air and water: the biology and physics of life's media*. Princeton University Press, Princeton NJ.

Heinrich, B. (1993). *The hot-blooded insects*. Harvard University Press, Cambridge, Mass.

Hokkanen, J. E. I. (1990). Temperature regulation of marine mammals. *Journal of Theoretical Biology* **145**, 465–85.

Monteith, J. L. and Unsworth, M. H. (1990). *Principles of environmental physics*, 2nd edn. Arnold, London.

Rome, L. C. and Bennett, A. F. (eds) (1990). Influence of temperature on muscle and locomotor performance. *American Journal of Physiology* **259**, R189–265.

Chapter 7

Bayne, B. L. and Widdows, J. (1978). The physiological ecology of two populations of *Mytilus edulis* L. *Oecologia* **37**, 137–62.

Bevan, R. M., Butler, P. J., Woakes, A. J., and Prince, P. A. (1995). The energy expenditure of free-ranging black-browed albatrosses. *Philosophical Transactions of the Royal Society* B **350**, 119–31.

Bryant, D. M. and Westerterp, K. R. (1980). The energy budget of the house martin (*Delichon urbica*). *Ardea* **68**, 91–102.

Daan, S., Deerenberg, C., and Dijkstra, C. (1996). Increased daily work precipitates natural death in the kestrel. *Journal of Animal Ecology* **65**, 539–44.

Hammond, K. A. and Diamond, J. (1997). Maximal sustained energy budgets in humans and animals. *Nature* **386**, 457–62.

Heinrich, B. (1979). *Bumblebee economics*. Harvard University Press, Cambridge, MA.

Nagy, K. A. (1987). Field metabolic rate and food requirement scaling in mammals and birds. *Ecological Monographs* **57**, 111–28.

O'Dor, R. K., Forsythe, J., Webber, D. M., Wells, J., and Wells, M. J. (1993). Activity levels of *Nautilus* in the wild. *Nature* **362**, 626–8.

Panter-Brick, C. (1993). Seasonality of energy expenditure during pregnancy and lactation for rural Nepali women. *American Journal of Clinical Nutrition* **57**, 620–8.

Priede, I. G. and Young, A. H. (1977). The ultrasonic telemetry of cardiac rhythms of wild brown trout (*Salmo trutta* L.) as an indicator of bio-energetics and behaviour. *Journal of Fish Biology* **10**, 299–318.

Thompson, D. and Fedak, M. A. (1993). Cardiac responses of grey seals during diving at sea. *Journal of Experimental Biology* **174**, 139–64.

Wolf, T. J. and Schmid-Hempel, P. (1989). Extra loads and foraging life span in honeybee workers. *Journal of Animal Ecology* **58**, 943–54.

Index